YORKSHIRE PLACENAMES

Dalesman Publishing Company Ltd
Stable Courtyard, Broughton Hall,
Skipton, North Yorkshire BD23 3AZ
www.dalesman.co.uk

First published 2001

Text © Peter Wright
Illustrations © John Ives

A British Library Cataloguing in Publication record
is available for this book

ISBN 1 85568 190 0

Cover photograph: the old North Riding sign at Thornton, near Middlesbrough
by Mike Kipling

Printed by Amadeus Press, Cleckheaton

YORKSHIRE PLACENAMES

PETER WRIGHT

Dalesman

ACKNOWLEDGEMENTS

To my family for their patience, to John G. Durnale for most timely background information and to all North and East Riding people who have so kindly assisted.

CONTENTS

Askrigg

INTRODUCTION

A long, detailed, historical introduction was never intended, not from disrespect to eminent historians but because history so far as this book is concerned lies very much in the word examples. Among the problems to overcome was how to divide the material. Yorkshire is easily the largest English county and, even without the solid West Riding, a great area had to be covered. The very rough plan adopted was to organise the basic picture in four parts: York area, western dales, the moors of North Yorkshire with Cleveland, and the East Coast plus the East Riding. To those were added other sections such as rivers and water, other countryside features like hills, mountains and dales, other places of interest, street names, and those whose pronunciation differed widely from spelling.

The plan was complicated by recent boundary changes and cannot please everyone: it is only a very approximate geographical guide. So, if the particular placename you seek is discussed, say, as belonging to the East Coast when you last saw it a few miles inland on the moors, or if what you treat as an entry into one of the western dales has been arbitrarily pushed into the York area, please be tolerant.

A minor difficulty is that quite often two or more places have the same name, as with the various Newtons and Daltons. Care is therefore needed to distinguish them geographically.

Experts in historical placenames depend a great deal on phonology, the development of different groups of words and sounds through the ages. By considering old spellings, they can very often see how the sound of a word has changed and can calculate or identify its origin. This can often be combined with knowledge of topography, the local geography, to avoid the most horrible mistakes.

The general pattern was to concentrate on tourist Yorkshire, which is chiefly the North and East Ridings, though that is not to deny that the West Riding has many places worth visiting. Yorkshire, as is well known, has long ago split into three Ridings (thirds, you might say, from Old Norse thrithjungr meaning 'a third part'). Ridings themselves were split into smaller units known as wapentakes, a Scandinavian word coming apparently from the brandishing of weapons at a council meeting to show agreement with the laws. However, to cut the placename material into such short sections would have made it unmanageable.

BE ON YOUR GUARD!

A placename seeker has to be constantly on guard against misleading words for, as in other aspects of life, things are not always as they seem. For example, Redmire west of Wensley is neither red nor a mire; it originally meant 'mere or lake (OE mere) where reeds (OE hreod) grew'. Booze, a small hamlet in Arkengarthdale, was not the very isolated haunt of heavy drinkers but from Bowehous meaning 'house by the bow or curve of land' (OE boga + hûs). Crackpot, four miles down the River Swale from Muker, is not named from a madman but means 'waterhole where crows abound' (probably linked with Swedish dialect pott and Old Norse krake).

Many errors of this nature arise from not knowing the name of the original owner. Thus Nunnington is not a village full of nuns but from an Old English personal name Nunna; Romanby next to Northallerton is unconnected with the Romans but merely 'village of Hrômundr', an Old Norse personal name; Wiganthorpe differs from Lancashire's Wigan in coming from a Scandinavian's very fitting name Vikingr;

Giggleswick was not from a man who could not keep his face straight but more normally the farm of one called Gikel; Menthorpe east of Selby was not a solely male community but perhaps slightly the opposite, commemorating a Scandinavian lady Menja. And what can be learnt from Fryup on the high moors near Danby? Whether or not the place has a fish-and-chip shop, which anyway seems unlikely, it probably means 'Frîga's small valley' from the Old English personal name + Old English hop 'valley'.

There are delightfully pleasant names like Appletreewick, Cherry Burton near Market Weighton, Belle View near Stokesley or, presumably best of all, Paradise, three miles west of Acklam. Such names contrast sharply with a highly repulsive one if its meaning is known, Urra on the North Yorks Moors from Old English horh 'filth' + haugr 'hill', or Bloody Vale off Swaledale from Old English blôdig. Equally objectionable appears Stank though it may not be so bad, Middle English stank apparently carrying the meanings 'pool, pond'.

Very fortunately many awkwardly sounding placenames have more pleasant meanings, often comparatively harmless, just from the names of their first owners. This, for example, should appease visitors to Sewerby Hall and gardens near Bridlington, the vital word meaning only 'Siward's farmstead' from a Scandinavian owner Sigworthr. Similarly Warthill near York is a praiseworthy name from Old Norse vartha and Old English hyll meaning 'beacon hill'. Sinnington folk near Pickering are not being branded as highly immoral, because where they dwell simply takes its name from the River Seven on which it stands. Or, if we suspiciously search through the East Riding, Bugthorpe is just 'Buggi's village' from a Scandinavian nickname, Spittle near Pocklington is just a shortened form of the word hospital, and Aike near Beverley stands merely for Old English âc 'oak-tree'.

Certainly some names do seem odd. Thus, for example, Snilesworth between the Cleveland and Hambleton Hills (Old

Norse snigill 'snail' and vath 'ford') and to the north-west Potto (probably from Old Norse pott 'waterhole' and Old English hoh 'hill'). For nearly all such a reason will be found, generally unearthing a less surprising name.

One thing that does seem surprising is that Yorkshire unlike other counties appears to have a smaller proportion of complaining names like Unthank or wildly remote ones like Jericho or Egypt implying 'away from normal civilisation'. This may suggest that, even if a Yorkshire 'Tyke' has to toil on rough, uncultivated land, without too much grumbling he will do his best, not a bad philosophy.

For general reading advice, there is a rhyme that goes:

'In ley and ham and hill and ton,
Many Old English placenames run,
But beck and kirk and by of course,
Arrive in Yorkshire from Old Norse'.

You can profitably amend or add to that rhyme. Many other placename elements appear on the following pages. Browse through them for interest or use them for checking at your leisure.

SPELLINGS

Old English and Old Norse had different characters or symbols for the voiced and unvoiced sounds we write th as in this and thin. For ease of reading we have kept to th and the letters around it will usually suggest whether it was voiced or not. Additionally, for 'short a' in modern Standard English, Old English used a joined a and e (known technically, if you are eager to know, as 'ash'); but again for simplicity we have kept them apart, as e.g. when referring to Old English aeppel 'apple' or blaec 'black'. You will notice this only occasionally.

ABBREVIATIONS

The only ones to note particularly are:

OE Old English (i.e. Anglo-Saxon), ON Old Norse (i.e. Old Scandinavian), OFr Old French.

ME Middle English

East Witton

THE YORK AREA

Acomb: Just west of York. From the OE dative plural âcum meaning 'at the oaks'.

Aldborough: 'Old stronghold' from OE (e) ald + burh. Originally a Roman settlement sprang up at a crossing of the River Ure. It was called Isurium Brigantium from a Brigantian name of the river. It was the chief town of the largest Roman tribe in Britain.

Aldwark: 'Old fortification' from OE (e) ald + OE (ge) weork.

Allerton Maulevere: Three miles east of Knaresborough. The first word means 'Aelfhere's farm' from the OE personal name, one like Alfred, and OE tûn 'settlement'. To this the Normans oddly added Maulevere 'bad harrier', though their reason for this is obscure.

Amotherly: 'Eymund's farm' from the ON personal name

Eymundr, changed in 12th century Norfolk records to Eimund.

Askham Bryan: Just south-west of York. First word from the dative plural OE or ON âskum + OE tûn, making 'village by the ash trees'.

Bagby: 'Baggi's farm' from an ON personal name Baggi + ON by.

Barton-le-Street: In the Malton area. Like Appleton-le-Street to the SE and farther away Wharram-le-Street it lies on an old Roman road. The name is a curious mixture from OE bere + tûn + French le + Latin strata (via) meaning 'barley farm by the paved road'.

Blubberhouses: Meaning doubtful. It has been thought to stand for 'black hill' or that it is connected with the foaming or bubbling of nearby Fewston Reservoir, though that does not seem to happen. It was certainly from an old dative plural as shown by 1172 spelling Bluberhusum equalling 'at the (whatever they were) houses'. Further suggestions welcome.

Boroughbridge: From OE burh + OE brycg to mean 'fortified bridge'.

Brandsby: North of Easingwold. 'Brand's farm' from ON personal name Brandr + by.

Butterwick: Hamlet on the River Rye in Pickering Vale. From OE wic + OE butere 'dairy farm with rich pastures'.

Cawood: Authorities disagree about this name. Possibly so called from the cawing of jack-daws + OE wudu 'wood', but more likely either 'cold wood' (first element from OE ceald) or from an Anglo-Saxon owner 'Ceolf's wood'.

Claxton: North-east of York. From an old Danish personal name Klak + tûn 'settlement'.

Coneysthorpe: Very near Castle Howard. Unconnected with rabbits (as Stanley Ellis in The Yorkshire Dialect Society Transactions for 1973 warns us), but meaning 'king's village' from ON kunung + thorp.

Copmanthorpe: South-east of York. ON kaupmanna + thorp 'outlying settlement of wandering pedlars', the first word

being connected with cheap and names like Cheapside in London.

Crayke: Near Easingwold. From Old Welsh creig, modern Welsh craig 'crag, rock'; and it does stand on a type of precipice.

I vividly remember Crayke through having to go there to help the training of a new Leeds University researcher who later became a professor of linguistics in Canada. Unfortunately my best informant was a very pub-lic-spirited elderly man who mis-guidedly thought I was seeking vulgar words and could produce a formidable variety. Consequently I feared what my younger colleague would make of this strange encounter, but in the end I was far more embar-rassed than he was! Crayke, only like many other Yorkshire localities, must have had its characters.

Dalby: East of Easingwold from ON dalr + by. 'Valley settle-ment'.

Duggleby: Near Malton. Domesday Book has Difgelibi 'Dufgall's Place', according to Morris a Norse name from Old Irish Dubhgall meaning 'black foreigner' and used in Ireland of Norwegian raiders who terrified coast-dwellers.

Earswick: 'Ethelric's abode' from that OE personal name + OE wic. Earswick used to be the northernmost outlier of York.

Easingwold: Many words with –ing are patronymics, denoting someone's descendants or followers, and Easingwold's meaning is 'forest land of Esa and his followers' from Esa + –ing + weald 'forest'. The Kentish Weald is fairly high, for example, but here the word is not confined to uplands.

Ferrensby: North of Harrogate. From ON faereyingr 'man or men from the Faroe Islands' + Old Norse by 'abode'. In the days of the old sagas, long before the arrival of modern Saga and other holidays, taking part in such a long voyage would make a seafarer famous in rural Danelaw, the territory controlled by Danes.

Flaxton: North-east of York. Some farms were named from their crops and here at Flaxton flax (OE fleax) was grown.

Foston: Also north-east of York. 'Fôtr's dwelling' from a mixed combination of the Scandinavian personal name and OE tûn. A t disappeared early from the spelling of the owner's name. The same etymology applies to Foston-on-the-Wolds in the East Riding.

Fulford: Now a southern suburb of York. The 1828 name was Fulfords Ambo, Ambo being Latin for 'both', because there were actually two villages making the same parish. Separately they were known as Water (OE waeter) Fulford, because it was on the River Ouse, and Gate Fulford, since it was near the gate (ON gata) or way carrying the York to Doncaster road over a small stream. Fulford means 'foul, dirty ford' (OE fûl + ford). Hopefully this unfortunately meaning to most residents is quite unknown, and deserves to be as modern drainage has vastly improved.

Gate Helmsley: On the old Roman road from York to Malton. Gate is from ON gata 'way, road' and the second word from OE personal name Hemele + either OE eg 'land' or leah 'enclosure'.

Gilling: A village that from the time of the old wapentake districts Gilling East and Gilling West has given its name to a gap between the Cleveland and the Howardian Hills. Probably from an OE personal name Getla because Bede, who finished his *History of the English Church and People* in AD 731, writes of Ingetlingum, reminiscent of the Getlingas, Anglian followers of Getla.

Goldsborough: Near Knaresborough. Domesday Book spelling Golborg comes from an OE personal name Golda. The second element burh 'fortified place' suggests it was an old military site.

Great Barugh: Between Malton and Pickering. 'Great mound or hill' from OE great + OE beorg or Scandinavian berg.

Habton – Great and Little: Near Malton. In 1086 at the time of Domesday Book Yorkshire and indeed England had only one Habbetun, 'Habba's settlement'. But in the next century arose a daughter hamlet, Parva 'Little' Habeton. Today, despite the rather grand name of one of them, they are not major dwelling areas.

Harrogate: Its original meaning was 'enclosed corner' from ON hagr 'hedge' + wraa a 'turn' + ON gata 'way, road'. Its Stray (Middle English straien 'to wander'), far more important than it sounds, is a wide central expanse of lawn very useful for dog walkers and all in need of relaxation.

Helperby: Near Easingwold. A suggestion has been 'Hjalp's farmstead' from a Scandinavian woman's name + by. Women

owning farms were not unknown in the 10th century but more likely from spellings is 'dwelling of Helpric'.

Hutton: Sand Hutton north-east of York is named from the sandy neighbourhood plus OE hoh 'spur of high ground like that in the name Plymouth Hoe' + tûn. Therefore the meaning of Sandy Hutton is 'dwelling(s) on a high sandy spur'. On the other hand Sheriff Hutton north of York takes its first name from a sheriff, OE scîr-gerefa or ME shire-reeve, an important city official, this particular person being Bertram de Bulmer who died in 1166.

Kelfield: South of York. 'Chalky field' from OE cealc + feld.

Kirkby Overblow: South-east of Harrogate. Overblow is a smelting term from steelmaking etc. One would expect it more around Sheffield but this area was once more industrialised. The meaning, coming from ON kirkja + by + Overblow, is 'settlement of smelters near the church'.

I'm afraid it is hard to believe the local legend that the Overblow was most helpful to a young lady who despairing of love threw herself off nearby Almes Cliff but luckily was saved by her billowing petticoats and skirts which supported a perfect parachute-type landing! The meaning of Overblow is hardly so romantic.

Kirkham: On the York-Malton road. 'Homestead with a church', but older spellings are too mixed to decide whether the name is from OE cirice + hâm or ON kirkja + heim.

Knaresborough: Originally it meant 'castle (OE burh) of Kenward or Cyneward' but, because it stands on a rocky slope, after 1250 it took note of the ME word knarre 'rough, hard, wooden knot' to change its supposed meaning to 'ford of the rugged rock'. The great medieval castle guarded the entrance to Nidderdale but few traces of it remain. Most prominent now are a keep, wall and wall towers built around 1350. Other well-visited sites in Knaresborough are of Mother Shipton's Cave and Petrifying Well where objects like kettles and old boots hang down from the cave top frozen by dripping water.

Langwith: Near Wheldrake. 'Long ford' from OE lang + ON vath, which sometimes survives in compounds as with.

Lilling: Ten miles NNE of York. 'People of Lilla' from the personal name + -ingas for his followers.

Linton-on-Ouse: West of York. From OE lin 'flax, linen cloth' because flax was grown there to make linen.

Malton: It has been suggested that it might mean 'malt settlement' from OE m(e)alt + tûn or, starting from OE mâl, 'tax, tribute town'; but most probably it arises from OE maethel 'speech, especially in a formal council' so that the underlying meaning could be 'discussion village', a kind of very local parliament.

Marton-in-the-Forest: North of York. ON marr 'fen, marsh' + tûn. Marton is fairly frequent in Yorkshire placenames.

Murton: East of York. Named from a natural feature. 'Farm on the moor' from OE môr + tûn.

Pocklington: Probably 'farm of Pocela's people' from Pocela + –ing + tûn. But another claim has been for 'town of Puccla" from OE pûcel 'goblin', seemingly linked with the fairy Puck who appears in Shakespeare's *Midsummer Night's Dream*.

Raskelf: Near Easingwold. Probably OE râ 'roe deer' + ON skjalf 'shelving, sloping ground'.

Rawcliffe: Near Overton on the Ouse just north-west of York. Its name refers to the reddish bank of the river there. It illustrates what must have happened to many an OE word, being replaced by a Scandinavian alternative, for here OE read 'red' has given way to ON rauthr of the same meaning before ON klif. The same etymology applies to the East Riding Rawcliffe west of Goole.

Ripley: Site of Ripley Castle, has a name meaning 'meadow (OE leâh) of Rippa' which may result from the clearing of woodland.

Ripon: Probably takes its name from Ripum 'on the banks' found in The OE Chronicles for 769 and ultimately Latin ripa 'bank', those here being the banks of the River Ure. In and around Ripon are a number of interesting tourist sites. Ripon itself has Ripon Cathedral with one of Europe's oldest crypts (built 672), Ripon workhouses in Allhallowsgate (OE âl + OE hâlig + ON gata 'road'), showing accommodation for tramps, and Old Sleningford Hall with an unusual garden and island watermill. Three miles south-west of Ripon is Fountains Abbey under the National Trust, the same distance south-east of it is historic Newby Hall and three miles north is Norton Conyers Hall, visited by Charlotte Bronte in 1839 and apparently the source of Thornfield Hall in her *Jane Eyre* because the hall had a legend resembling the story of the mad woman in her novel.

Rillington: A placename of dubious history. In Domesday Book it is spelt Redlington and Renliton, but by 1391 the spelling is Rillington. Originally it may have been Hredle + tûn, 'Hredle's headquarters'.

Slingsby: West of Malton. H.G.Stokes believed it meant the abode of an idler but it seems safer to derive it from a Scandinavian personal name Slengr to mean 'Sleng's farm'. In Domesday Book Sleng- forms outnumber Sling- ones.

Spofforth: In Domesday Book spelt Spofod. Spot was the recorded name so it might mean 'ford of Spot'. Spofforth Castle was once owned by the Percy family. Now its ruins are easily accessible, entry free. I remember the village especially from recording an elderly farm worker who had been blind from middle age but still managed to garden by means of white sticks connected by thread, illustrating old-time difficulties and determination in meeting them.

Stillington: North of York. A patronymic again, referring to followers. From OE Styfela, the owner, + ing for his followers + tûn. Thus 'dwelling of Styfela's people'.

Stockton-on-the-Forest:
North-east of York. 'Stockaded enclosure' from OE stocc 'tree stump' + tûn, with The Forest being The Forest of Galtree. A small but curious matter about this name is the use of the preposition on instead of in, as could be said e.g. of a disabled aircraft landing tragically on a forest.

Storthwaite: South-east of York. 'Bullock field' from OE styric, dialect stirk, +.ON thveit.

Storwood: 'Brushwood clearing', from ON storth + ON thveit.

Strensall: 'Streon's nook of land' from the personal name + OE healh usually meaning 'nook, corner'. In modern times well-known to the Army.

Swinton: Near Malton. 'Pig farm' from OE swin + tûn, having the same etymology as another Swinton 9m north-west of Ripon. By no means an uncommon placename, as many places must have been most obvious from the animals kept there.

Thicket: Next to the River Derwent near Wheldrake. 'Thick head'. Not slang but from OE thicce + heafod, meaning here in more detail 'river headland with thick vegetation'.

Towthorpe: 'Tove's village'

from ON Tôfi + thorp. A Towthorpe with a similar language history is on the Yorkshire Wolds (see East Riding Section).

Warthill: Near York (probably 'beacon hill' from ON vartha + OE hyll), it is ready to show discerning visitors its Brookfield Hall.

Weeton: North of Leeds. 'Willow farm' from OE withig + tûn. An East Riding Weeton near the tip of Holderness has the same word history.

Whenby: North of York. Spelt Quennebi in Domesday Book. From ON kvenna 'of the women' + ON by, so meaning 'women's settlement'. This suggests it was not rare for women as well as men to give their names to villages, hamlets and pieces of land brought into cultivation from scrub or forest.

York: Although this city's old name Eoforwic may seem very Anglo-Saxon and as if it meant 'boar territory', appearances can be deceptive. In AD 71 the Romans built there an important fortress which for the next 400 years was known as Eboracum, which became under the Angles Eoforwic. Even today the Archbishop of York signs for himself, as is York Minster's custom, in Latin Ebor. Perhaps baffled by the Eoforwic, which closely resembled their own word eofor 'wild boar', citizens changed the name gradually via spellings such as Euric. The modern pronunciation comes from the Danes, who called it Jorvic (their J equalling our Y), from which at last we get the city's name York.

Of York suburbs the strangest-looking is **Dringhouses** on its south-west boundary, recorded in the 13th century as Drengus and Drenghous. It stems from ON drengr, equal to OE dreng and applied first to a young person but later to a privileged peasant holding his land by old pre-Conquest tenure. Although the suburb's name is very little altered, it would be most surprising if current tenants in that part of Yorkshire's most central if not capital city depended on such ancient documents!

Grinton

THE DALES

Here are treated the western and north-western dales of the North Riding as opposed to any in Cleveland or the East Riding.

Ainderby: Means 'dwelling or village of Eindrithi', a Scandinavian. Domesday Book has the spelling Eindrebi. Ainderby Steeple, west of Northallerton, is so named from its prominent church tower. Ainderby Quernhow, west of Thirsk, is associated with milling, shown by its origin from ON kvern 'millstone' and haugr 'mound'; the third, Ainderby Mires, receives its name from the swampy terrain.

Aiskew: Next to Bedale. From ON eikr + skôgr, 'oak wood'.

Aislaby: In Eskdale. ON personal name Âslâhr + by, meaning "Âslâhr's farm'.

Appersett: Very close to Hawes in Wensleydale. From OE aeppeltreow + ON saetr. The meaning is 'mountain pasture by the apple tree'.

Arncliffe: In Wharfedale. 'Eagles' cliff' from OE earn + clif. The Rev. Charles Kingsley stayed at a house near the bridge and used the area as background for his book *The Water Babies*, a protest much needed in those days against the exploitation of children by making them climb up sooty chimneys to sweep them.

Askrigg: In Wensleydale. In Domesday Book it is spelt Ascric. It means 'the ash-tree ridge', from OE aesc and ON hryggr.

Aysgarth: Also in Wensleydale. ON eikr 'oak' and ON skarth 'wooded clearing' combine to give its full translation, 'open space marked by oak trees'. Aysgarth was in the middle of the Forest of Wensleydale and its wooded nature is indicated by various placenames, such as the thwaite 'clearing' in the name Swinithwaite three miles down the river.

Bainbridge: In Upper Wensleydale. It is named from a bridge over the River Bain, a river meaning 'straight', where it has replaced a fort where that little river from Semerwater joins the Ure. At many such river crossings, especially the well-used ones, a ford gave way to a bridge.

Bedale: Despite various other interpretations, such being from OE be dal 'by, near the dale, or even 'bee dale' from OE beo, it is probably from OE Beda + healh 'Bede's corner of land'.

Bellerby: Near Leyburn in Wensleydale. 'Belg's farm' from an ON name Belgr + by, the personal name apparently meaning at first 'bellows' and then 'withered old man'.

Bentham – High and Low:
'Home among the bennet, rush grass', from OE beonet + hâm. Compare Chequerbent and Chowbent near Wigan.

Buckden: On the Upper Wharfe below Buckden Pike. Probably from OE bucc 'stag' + OE denu 'valley'.

Burnsall: Near Grassington. OE burna 'stream' + OE h(e)alh 'corner', i.e. alluvial land. In wilder parts of Northern England such riverside pastures were particularly important.

Carperby: In Wensleydale. Old Irish Cairpre, a personal name meaning charioteer.

Catterick: In Roman times there was a fort at what they called Cataractonium. Latin cataracta means 'waterfall' and near this point the River Swale does flow very fast. There is probably a connection with Welsh cader 'hill fort'. (Compare Cader Idris, rather forbidding if you attempt that climb in mist.) Seemingly Catterick was the only partly British-named settlement in the North Riding.

Cleasby: Near Darlington. Probably 'dwelling of Clea or someone with a similar name'.

Cockleberry: East of Scotch Corner. Probably OE coccel 'cockle' + OE be(o)rg 'hill'.

Colburn: Near Richmond. Probably 'cool stream' from OE côl + burna rather than from OE col, cf. ON kol 'coal', from the dark colour of the water.

Coldknuckle: North-east of Melsonby. Self-explanatory; warmth against the elements seemingly vital.

Constable Burton: Near Leyburn. First recorded as plain Burton from OE burh + tûn, meaning 'fortified farmstead' but after the Earl of Richmond granted the manor to his chief officer about 1100, there was added the Norman title

Constable.

Countersett: Near Semerwater off Wensleydale. From Norwegian saetr, an upland pasture or huts on it from which shepherds and herdsmen spent their summers looking after their animals. In the 13th century to this place was added the Old French personal name Constance to give the spelling Constansate.

Crackpot: In Swaledale. No slangy reflection on the local intelligence, but 'pot where crows abound'. First element probably ON kraka + pot, an old Cumbrian word for a deep lake or pool (compare Swedish dialect pott 'water hole'). The pot would be caused by a rift in the limestone.

Dalton: 'Valley farm', from OE dael + tûn. It is hardly surprising that Yorkshire with so many dales has a number of Daltons, all with the same etymology. Around the Dalton NE of Richmond in the 12th and 13th centuries three were mentioned, namely Dalton Trauers, Dalton Mitchell and Dalton Norreys named from local tenants, besides which we have today Dalton south of Thirsk, North Dalton, east of Pocklington, Dalton-on-Tees, South Dalton in the Beverley area, etc. When therefore writing to someone in a Dalton, do please remember to include your postcode.

Ellerton: Near Catterick. 'Alder enclosure' from OE alor (ON elri) + tûn.

Gargrave: Has probably two Anglian elements in its name – gara 'triangular piece of land left by ploughing' and grâf 'grove'. I recollect Gargrate from being grounded there in dialect search for the inside of a week with our university 'dialect car' in a garage with cylinders refusing to fire after obstinate travel over the mountains from Hawes, in one steep and winding road section having to zigzag upwards in first gear, having ignored sensible advice to take a much longer but level route via the A1. However, all the blame was defensively put on the North Riding mountains!

Grassington: 'Town of Gersent or Gersendis'. Very picturesque, with sloping main street and cobbled square. I remember those cobbles before they were removed in 1965 and later restored, and again from dialect hunting through having to be towed again and again round that square, then deep in snow, till the tow rope broke.

Greenhow Hill: In Nidderdale. Its name includes how 'mound' from ON haugr. Despite its name it is a village and, though lower than, say, Flash in Derbyshire, it can claim to be one of England's loftiest villages.

Grinton: In Swaledale. 'Green enclosure' from OE grene + tûn.

Gunnerside: In Upper Swaledale. ON personal name Gunnr + saetre 'upland pasture'. In both Yorkshire and Cumbria such occurrences of saetr meant both sloping uplands and their huts from which shepherds and herdsmen could supervise their animals.

Hartforth: Near Richmond and close to the section of road north-west from Scotch Corner formerly called Watling Street. The meaning 'hart ford' is from OE heorot + ford. Many place-names are from animals using them or found near them.

Hartwith: In Nidderdale. 'Hart wood', named from the fallow deer, so another animal name like the last item. A hybrid name from OE heorot + ON vîthr.

Hawes: In Wensleydale. The meaning is 'a pass between mountains' from OE and ON hals 'neck', with the l having changed to a 'short u' as can happen historically, i.e. in strong Cockney. Even in 1928 A.H.Smith, well-known delver into placenames, was reporting 'the village is only of recent

growth but for a long time now it has continued expanding and is a great favourite of tourists.' Known for the efforts of benefactor Kit Calvert in the past and others today in cheesemaking etc.

Hellifield: Either 'Helgi's field', 'Halga's field', or 'holy field' from OE haelig + feld. The first alternative may look preferable, but to help you decide the Doomsday Book spellings are Helg- and Haelgefeld.

Hipswell: Near Richmond. 'Well at the hipple, little heap' from OE hypothetical hiepel. There is a lone hill west of the village.

Horton-in-Ribblesdale: Ribblesdale of course indicates the nearness to Lancashire with its River Ribble, probably from an OE rîpel 'tearing'. Horton is more doubtful, perhaps meaning an area specialised in horse rearing and certainly a pleasanter derivation, as its residents will agree, than an alternative put forward, OE horh tîn, 'dirty, muddy town'.

Hudswell: In Swaledale. By contrast comparatively straightforward in word development, from an early spelling Hud(e)leswell to mean 'Hudd's well' from an OE personal name Hudd + w(i)ella.

Hutton Conyers: Near Ripon.

'Farm on a spur of land' from OE hoh + tûn + Conyers added from a 12th century family that acquired land nearby. Its castle, apparently built to extort tribute from subservient Ripon townsfolk, was demolished by Henry II.

Hutton Hang: Lies interestingly just south of Finghall (seemingly connected with ON thing 'assembly') in Wensleydale. Its name comes from Hutton (See preceding item). Hang from Hang Bank, a hill hanging just north of it which was apparently a wapentake political meeting place.

Ingleton: 'Abode of the Angle, Northern Englishman', from OE Engle + tûn. Notable for its caves, especially White Scar (OE hwît + perhaps OFr escare) Cave on the road towards Chapel-le-Dale.

Keld: A hamlet next to Muker at the head of Swaledale, from ON kelda 'spring'. So important was water for early settlers that many placenames are formed from this word. In the 14th century Keld had a longer name, Appeltrekelde, first element from OE aeppeltreôw and meaning 'spring near the apple tree'. The hamlet is etched in my memory because in the 1950s I had to motor in Keld over the stripling River Swale and up a fearsome S-bend road with a gradient officially signposted 1 in 2. Nowadays it is next to the Pennine Way. Descending, necessary more than once because of different appointments, was particularly hair-raising as, even in first gear and with brakes full on, the car could not quite stop. Presumably since then the transport authorities must have noticed and eased the climb.

Linton

Kepwick: Near Northallerton. 'Market place' from OE ceâp 'price, bargain' + wic 'place'.

Kettlewell: A fine rambling area on the Upper Wharfe below Great Whernside. It apparently means 'spring or stream in a narrow valley' from OE cetel + wella, with Scandinavian K- replacing English Ch- of the former name Chettlewell.

Kiplin: Near Catterick. 'Settlement of the Cippelings' with loss of the plural 's' from OE personal name Cippa, found also in Chippenham, Wiltshire.

Kirklington: On the lower Ure eight miles west of Thirsk. 'Farm of the relatives of Cyrtla'. From OE Cyrtla + ing 'followers' + tûn. The change of sound t to k before l easily happens as when children say likkle for little.

Lastingham: North of Northallerton. Another family-group name from OE personal name Last + ing 'followers' + ham. 'Village of Last and his dependants'.

Leeming: Near Bedale. Originally from the river name from an old British word parallel to Old Irish leamh 'elm'. 'River by the elm trees'.

Leyburn: In Wensleydale. 'Stream by the forest clearing' from OE leah + burna.

Litherskew: Near Aysgarth. 'Hillside wood' from ON hlith 'slope' + ON skôgr 'wood'.

Linton: Next to Grassington. 'Settlement where flax was grown' from OE lîn 'flax' + tûn. Linton is a charming village with waterfalls, stepping stones and one of the oldest churches in the Dales.

Masham: On the Lower Ure. 'Massa's homestead' from OE Maessa + hâm.

Melsonby: North of Scotch Corner. From Old Irish to OE Maelsuthan, name of a financier to Edward II, + ON by 'village'.

Mickleton: Near Middleton-in Teesdale. 'Large farm' from OE mycel 'big' + tûn. Compare the Yorkshire saying, 'Every mickle makes a muckle'.

Middleham: Near Leyburn, simply describing its situation. 'Middle settlement' from OE middel + tûn.

Moreton-upon-Swale: 'Farm on the moor' from OE môr + tûn.

Moulton: Near Scotch Corner. 'Mula's farm' from OE Mûla + tûn.

Muker: In Swaledale, 'small cultivated field' from ON mjôr 'thin, narrow' + acre 'small field', source of our modern measurement, the acre.

It was at a Muker inn, where in

1950 Professor Harold Orton and I stayed, that, watching shepherds play dominoes, we might have been the last people in England apart from those players to hear genuine use of the old sheep-scoring numerals. On another occasion, having lost my way, I was stopped high above Muker in a poultry trap. Getting out of the car to explain, I could hardly keep my feet on the ice. Dwellers in that area in the depths of winter deserve our utmost sympathy.

Myton-on-Swale: 'Farm where the rivers meet' from OE mûth + tûn. It was the site of a battle in 1319 mentioned by Barbour.

Nesfield: Upsteam from Ilkley. 'Open land on which cattle are kept'. From OE neates 'beasts' + feld 'field'.

Newbiggin: In Wensleydale. 'New building'. From OE nîwe 'new' + bigging 'building' from ON byggja 'to build'.

Newton-le-Willows: Near Leyburn. 'New farm' from OE nîwe + tûn, later distinguished by the help of French le 'the' and OE wilig as 'where willow trees abound'.

Newton Morrell: South-west of Darlington. Newton appears again, but this time with the added name of a family of French landowners there.

Nosterfield: Near Ripon. At first glance it looks affected by Latin noster 'our' as in pater-noster; but actually it comes from mistakenly joining a last letter n of a previous word to OE eowerstrefelda 'the ewes or sheep fold'. It serves to remind us that farming is not only our oldest industry but once was our largest.

Pateley Bridge: An old market town, comes from OE paethleah 'path by a meadow' + OE brycg 'bridge'.

Reeth: An old market centre in Wensleydale with a large and curiously sloping village green. The name is from OE rithe 'stream' which may well be the Arkle Beck flowing through Reeth, whose pronunciation despite spelling changes has been basically unaltered since the times of the Anglo-Saxons.

Richmond: In Domesday Book it was called Hindrelac (origin unclear) but after the Norman Conquest it gained the name Richemund 'strong hill', all or partly French from OFr riche plus OE munt 'mountain' (spelt in 1176 Richemunt) and ME and Latin mont of similar meaning. Richmond Castle, topping a lofty precipice, was built by the Breton Alan on land granted him by William I. Also the substantial remains of Easby Abbey stand by the River Swale close to the town.

Richmondshire: Old name for land within the jurisdiction of the manor of Richmond.

Romaldkirk: From ON kirkja 'church' and the name of the saint to whom it is dedicated.

Scotch Corner: Four miles north-east of Richmond. Not in older placename books but useful for motorists to note the self-explanatory name as it starts the route to Penrith, Carlisle and Scotland beyond.

Scruton: In Swaledale, from an ON personal name + tûn. 'Skurfa's farm'.

Settle: In Domesday Book spelt Setel and from OE setl 'seat, resting place'. One of very few reasonably large places well within Yorkshire dales.

Skipton-in-Craven: Or just Skipton as it is generally known, being much larger than its namesake Skipton-upon-Swale. Both Skiptons, however, derive from Old Northumbrian scip 'sheep' + tûn. Skipton Castle is a massive one which has covered all signs of its Norman original. The very wide high street has a vigorous market area.

Studley Roger, Studley Royal: Near Ripon, the latter with a very interesting church. Both

places are reminders of horse-breeding, as Studley means 'meadow of the breeding stud' from OE or ON stôd + OE leah.

Swinton: On the River Ure near Masham. 'Pig farm' from OE swîne +tûn, with the same word development as e.g. another Swinton near Malton.

Tan Hill: Near Thwaite (ON thveit 'clearing'), famous as the highest inn in England. Height 1732 feet (528 metres), beating its Lake District rival at the top of the Kirkstone Pass by some 250 feet. The Pennine Way passes it among peat and bog.

Thornton Watless: Thornton is from an OE personal name Thorn or else OE thorn 'thorn tree'. Watless is of more doubtful origin. It has been believed to mean 'waterless' from ON vatn-lauss, but there is no water shortage here, between the River Ure and Bedale. More likely it means 'water clearings or meadows' from OE waeter + OE leâh, making the full meaning probably 'thorn trees by the water meadows'.

Thorpe: Near Grassington. It is a common Scandinavian place-name element meaning 'small outlying farm or hamlet', though it is rare to find it alone as here.

Threshfield: Perhaps 'field

where threshing took place'
from OE therscan + feld,
although an origin from Old
Swedish thraesk 'fen, lake' has
also been claimed.

Thrintoft: In Swaledale.
'Homestead near thorn bushes'
from OE thrynne + Danish topt.

Thwaite: Also in Swaledale,
from ON thveit 'meadow, clearing' with no extra word to identify it. In this it resembles
Grassington's neighbour Thorpe
above in standing alone.
Thwaite does not openly boast
of being the only thwaite worth
calling a thwaite as standing
alone might imply, though it
used to be called Arkeltwaite
'Arkel's clearing'.

Walburn: In Swaledale, probably meaning 'abode of foreigners by the stream' from OE
weala + OE burna.

Walden: In Wensleydale, 'valley
of foreigners' from OE weala
'foreigners' + OE denu 'valley'.
Names like this and the one
above may suggest that foreigners, e.g. Scandinavian settlers,
were heavily outnumbered, but
the evidence should not be
relied on unduly.

Walshford: In Lower
Nidderdale. Another placename
similar in origin to the last two.
First recorded in the 13th century as Walesford from OE weala

+ OE ford, 'ford of the
Welshman'.

Wensley: 'Waendel's forest
clearing' from OE personal
name Waendel + OW leah.
Once so flourishing a settlement
that it gave its name to
Wensleydale but the 16th century plague badly affected it
and now it seems rather off the
best-known tourist haunts.

Whaw: Beyond Reeth at the
very top (or head as the experts
say) of Arkengarthdale. From
OE haga or corresponding ON
hagi 'enclosure'. In Whaw it
was for cows.

Whitewell: East of Catterick.
'White, clear well' from OE hwît
+ OE w(i)ella.

Woodale: In Coverdale. The
meaning is 'wolfdale' from OE
wulf + OE dael, probably from
wolves prowling around.

Wycliffe: Overlooking the River
Tees east of Barnard Castle.
Probably 'white cliff' – there is
one there – from OE hwît + OE
clif.

Yafforth: Where the road from
Northallerton to Richmond
crosses the River Wiske.
Previously, as in so many places,
crossing was by means of a
ford. Domesday Book has
Iaforde 'ford across the river'
from OE ea 'river' + ford.

John A. Ives '92.

Helmsley

NORTH OF YORK AND CLEVELAND

Ampleforth: 'Ford where sorrel grows'. From OE ampre 'dock, sorrel' + ford, and in 1086 Domesday Book written Ampreford. It has probably England's best-known and deservedly famous Roman Catholic boys' school.

Antofts: In Ryedale. 'Aldwine's farmstead', from the OE personal name + Old Danish toft.

Appleton-le-Moors: Near Kirkbymoorside, from OE aeppel 'apple' + tûn + môr 'moor' because of its nearness to them.

Appleton-le-Street: Which the above Appleton faces, is 9 miles south near Malton. The Street in its name derives from a supposed Roman road on which it and nearby Barton-le-Street are thought to stand.

Battersby: Three miles southeast of Great Ayton, 'Bothvar's farmstead' from the ON personal name Bothvar + by.

Birkby: Six miles ENE of Northallerton. 'The Britons' village' from ON Breta-byr. The Domesday Book spelling is Bretebi.

Borrowby: On the A19 west of the Hambleton Hills. 'Hill farm' from OE be(o)rg + by.

In Borrowby, after a marvellous tea on newspaper for a table-cloth but no worse for that, a very kind and helpful farmer was watching me sketch his farm cart to illustrate his words for its parts when he began to believe I was anxious to collect old things rather than mere words and insisted on presenting me with an ancient rabbit gun. This I forgot to throw away as intended in a ditch on my route back to base, a hall chiefly for overseas students in Bramhope, but left it by my bed and the next day the maid, petrified to see it, called the police. Even a peaceful country trip can bring the unexpected.

Brompton: Near Northallerton and another halfway from Pickering to Scarborough have similar word developments meaning 'piece of land overgrown with broom, gorse' from OE brôm + tûn.

Carthorpe: Six miles west of Thirsk. 'Kari's village' from the ON personal name + thorp. Thorp is a Danish word and we meet many Thorps.

Cawton: Two miles WNW of Hovingham. 'Calves' farm' from OE calf + tûn. Another out of many placenames from animals kept on farms.

Chop Gate: The 14 miles or so of Bilsdale support very few people and this is the only village along that length. Pronounced by natives Chop Yat since gates (from ON gata 'road, way') are yats in the local dialect. Chop may be connected with Scandinavian kaup 'pedlar or chapman' for in Middle English times there were many tracks across The Cleveland Hills.

Cold Kirby: Most suitably named, high up near Sutton Bank in the Hambleton Hills where in winter it will catch the most biting winds. From OE cald or ON kaldr 'cold' + ON kirkja 'church' + ON by 'settlement'.

Coxwold: South-east of Thirsk. Coxwold's language history is significantly shown in an 8th century charter calling it Cuha-

walda 'Cuha's woodland'. It is a charming village popular with visitors because of its interesting church and Shandy Hall, where lived from 1760 to his death eight years later curate Lawrence Sterne, writer of the comic novel *Tristram Shandy* from which Shandy Hall, built as a timber-framed house in the 15th century, takes its name. Unlike many museums it is arranged as if he had just left it, with apparently even the sink unaltered. Along with two or three other literary figures like Fielding and Richardson, Lawrence Sterne was a forerunner of the modern English novel.

Danby: 'Village of the Danes'. The various Yorkshire names with Danby emphasise the strength of the Danish invasion and settlement of parts of the county. There is a Danby in the Esk Valley, the hamlet of Danby Wiske near Northallerton part-named from the River Wiske, along with others like Little Danby and Danby-on-Ure which have dropped away from some maps. No real shortage of Danbys.

Easby: On the North Yorkshire Moors it has a name hardly changed from Domesday Book Esebi. It stands for 'Esi's farm' from the Old Scandinavian personal name.

Egton: High above its twin village Egton Bridge down in the valley. Probably 'Ecga's farm' from an Old English personal name + tûn.

Everley: On the moors northwest of Scarborough. 'Wild boar clearing' from OE eofor + leah.

Faceby: South-west of Stokesley. 'Feit's farm' from first element ON feitr 'fat' used as a personal name.

Fadmoor: 500 feet up on the Cleveland slope above Kirkbymoorside. The Fad first part is probably from an unknown personal name followed by OE môr 'moor'.

Felixkirk: Near Thirsk. Some placenames have religious associations, as here ON kirkja 'church' plus St Felix, the name of the saint to whom it was dedicated. Its old name, however, was Fridebi 'Frithi's farm'.

Foxton: Near Northallerton, involving another wild animal name. 'Lair of the fox' from OE fox + tûn.

Fylingthorpe: 'Outlying settlement of the people of Fygela'. From an OE personal name + Scandinavian thorp. In the 13th century it was sometimes called Prestethorpe 'priest (OE preôst) village' because the land was held by the monks of Whitby.

Goathland: Theories have been put forward to link it with goats or a perfect place, 'God's land', but more probably it is simply 'Gôda's land' from an OE personal name.

Great Ayton: Ayton means 'farm by the river', from OE eâ 'river' + tûn. The Teesside saying canny Ayton probably springs from its picturesque situation. In Domesday Book it is spelt Atun and, as shown by a Whitby Abbey document, its daughter settlement Parva Hatona, now called Little Ayton, did not appear till a century later.

Grosmont: At the northern end of the North Yorkshire Moors railway from Pickering, though a continuation to Whitby has been proposed. It means 'big hill' from the French name for the mother priory at Limoges.

Hawnby: In Ryedale. 'Halmi's farm' from an Old Scanidavian by-name. The change in old spellings from an l to u and w resembles what took place in Anglo-Norman and is similar to the speech of those Cockneys who, instead of what used to be called a 'dark' or throaty l pronounce a 'short u'.

Helmsley: 'Hemel's forest clearing' from the OE personal name

+ OE leâh. Compare Gate Helmsley in this book's York section.

Hilton: Self-explanatory, OE hyll 'hill' + OE tûn 'settlement'. For residents doubtless less expensive and far more convenient than the Hilton Hotel.

Hole of Horcum: At the head of a rather deep valley. Rather puzzling to motorists on the moorland road between Pickering and Whitby. From OE horh 'filth, mud' + Celtic cumb 'valley'. The horh element is now out-of-date through field drainage and a rather repetitive Hole of has been added. According to legend, the giant Wade scooped out earth to make it and threw away the spare soil to create Blakey or Roseberry Topping. A similar story involving a giant centres on the making of the Wrekin in Shropshire.

Hornby: South-west of Yarm, like another Hornby north-west of Bedale, may be shortening of a Scandinavian name Hornbothi.

Huthwaite: North-east of Osmotherley. Since it is on a long narrow ridge overlooking Crook Beck, for derivation it seems safest to take OE hoh 'spur of land' + ON thwait.

Hutton: Basically 'high settlement'. The North Riding has

been found by research to have at least 18 places containing this name.

Hutton-le-Hole: 'Farm on projecting piece of land' from OE hoh 'spur of land' + tûn. The additional le Hole is simply from French le 'the' + OE hol 'hollow', giving us 'in the hollow'. A pretty village attracting more and more visitors. When I last went there, its policing looked old-fashioned but effective; an official had apparently put a notice in the front window of his terraced house warning visitors of a certain gentleman.

Hutton Rudby: Is south-west of Stokesley on one side of the River Leven with plain Rudby on the other. The latter name is all Scandinavian, 'Rudi's farm' from a rare ON personal name + by; whereas Hutton Rudby is half English, from hoh 'spur of land' + tûn. If there had been extreme racial hatred between them, the names of these places so adjacent to each other would not have persisted for over 1,000 years till today.

Ingleby: 'Village of the English' from OE Engle + Scandinavian by. Around Stokesley and all less than 10 miles from each other are to the north-west Ingleby Barwick (second element 'barley farm'), south-east of it Ingleby Greenhow (second word formed from OE grene + ON haugr 'hill') and to the south-west Ingleby Arncliffe (second word 'eagles' cliff' from earn + clif). They all probably denote isolated survivals of English natives in a predominantly Scandinavian population.

Jolby: Near Croft. 'Joel's farm' from OFr personal name Johel + by. The placename was first recorded in 1193-9 and the by with which it ended is the only clear proof of Scandinavian by 'settlement' being used as a living word to form a placename in the North Riding since the Norman Conquest. What a pity it does not appear on recent maps! Has its land been taken over for other purposes?

Kilton Thorpe: Near Loftus. Hard to decide whether Kilton is of OE or ON origin, but Thorpe shows it was a secondary outlying settlement depending on nearby Kilton.

Kirby in Cleveland: 'Farm by the church' from ON kirkja + by, with the addition in Cleveland stressing it is in a steep district just as occurs with Carlton-in-Cleveland. The in-Cleveland certainly adds distinction, and hopefully does not worry any visitor too much about comparative height.

Kirby Knowle: Near Thirsk and on the very edge of the Hambleton Hills. From Scandinavian kirkja + by as in the last example + Scandinavian knoll 'round-topped hill'.

Kirkbymoorside: Again from ON kirkja + by + the self-explanatory moorside. The 1399 ending to the name was more-sheved 'head, i.e. end of the moor'.

Kirkleatham: From an Old Norse dative plural hlithum 'at the slopes'.

Knayton: An Anglo-Saxon woman's name Cengifu + tûn. Men did not control everything!

Loftus: From ON lopthûsum 'at the house with lofts'. In some languages the lip sounds p and f have close links with the other of their pair in another language. Compare e.g. German pfennig and English penny, or as here the ON lopt with English loft.

Low and High Worsall: South-west of Yarm. Worsall comes from an OE personal name Wyrc + h(e)alh 'corner of land', i.e. near the River Tees. Low and High of course contrast the two neighbouring villages.

Middlesbrough: From OE personal name Midele + OE burh 'fortified place'. Beware of the spelling because, unlike most boroughs, it seems to have deliberately discarded an o.

Moorsholm: Near Guisborough. From Domesday Book Morehusum 'at the houses on the moor'.

Normanby: No connection with the Normans! Two Normanbys can be found in the general area of Cleveland and its surrounds, one now a suburb of Middlesbrough and the other near Pickering. Both mean 'village of the Norwegians' from OE Northman + by.

Northallerton: 'Alfhere's farm'. North was added to spellings for the place in 1371.

Nunthorpe: Now an outlier of constantly expanding Middlesbrough. In Domesday Book it was just Torp but in 1301 it became Nunnethorpe after the nuns of St James's Church, which was previously there.

Osmotherley: Probably 'Osmund's clearing' from the ON personal name Âsmundr + OE leâh.

Oswaldkirk: South of Helmsley, with kirk (ON kirkja) as so often denoting a Scandinavian settlement. 'Church dedicated to St Oswald'.

Picton: Three miles south of Yarm. From OE personal name Pîca (notice how many place names are of this personal

name type) + tûn. 'Pica's farm'.

Pockley: Near Helmsley from OE Poca + leâh. 'Poca's forest clearing'.

Redcar: From OE hreôd + ON kjarr. 'Reeds on marshy land'.

Salton: Near Pickering. OE s(e)ala + tûn. 'Enclosure by the willow trees'.

Sawdon: Near Beedale (not to be confused with Bedale!). Again OE s(e)ala + OE denu 'valley'. Thus 'willow valley'.

Scaling: Four miles from the North Yorkshire Moors centre at Danby. It means 'shieling, a rough wooden hut or temporary shelter'. From a conjectured ON skâling.

Sigston: Near Northallerton. Probably 'Sigg's farm' from ON Siggr + the very familiar OE tûn. Surname of one of my best informants in Cleveland, though on dialect quests research rules and respect for privacy fobade us to publish actual names, etc.

Skelton: Of doubtful etymology, recorded in Domesday Book as Chilton, which is hard to explain.

I remember Skelton most not so much for etymologies or pronunciations but for some pre-home-watch caution which apprehended me as a burglar on a preliminary search. In dialect collecting the names of things in a house including parts of a door had to be sought without giving away any likely dialect answer, and who better to provide all those household words than a housewife born and bred there? Thus I was in true academic fashion innocently seeking words for door jambs, snecks, apses or asps, etc. and for accuracy actually drawing one or two until the good lady of the house became suspicious and brought in neighbours to vet me. Full marks to Skelton for its wise attention to security.

Skiplam: Near Kirkbymoorside. 'At the cowsheds' (dialectal shippons) from OE scypen with Scandinavian sk- substituted for the softer sound we write sh but which was spelt in OE sc-.

Smeaton: South-east of Dalton -on-Tees. 'The smith's farm' from OE smith + tûn. Some settlements were named from groups of people there, in this case blacksmiths. In former times almost every village had its smith, which is why so many folk have the surname Smith today.

Sproxton: Near Helmsley. ON personal name Sprok, probably meaning 'brittle' + tûn.

Stokesley: Second element from OE leâh 'meadow' but the first very problematical.

West Burton

*Stokesley's 'foreign' element
was once a problem for me in
another sense for, returning
from there delighted with what
seemed an accurate array of
local words, pronunciations and
grammar, I learnt that one of
my willing helpers was not technically a native but had spent
some of his infant years in
Sussex and so I had to go all the
way back to question another
genuine Stokesleyite.*

Thirkleby: Near Kilburn.
'Thorkel's farmstead' from ON
Thorkell + by. A similarly devel-
oped Thirkleby honours
Holderness (see that section).

Thirsk: Especially as it is in a
well-watered area, is most likely
from ON thraesk 'fen, lake'. It is
near districts very well-known to
tourists, particularly as it was
from there, in one of the main
streets, that Alf White, TV's
James Herriot, organised his
practice.

Thornaby-on-Tees:
'Thormoth's farm' from the ON
personal name Thormothr + by.
However, to be ridiculously
pedantic, not actually on the
Tees, which would be uncom-
fortable, but next to it.

Thornton-Dale: Near Pickering
translates as 'enclosure made by
thorn bushes', from ON thorn +
tûn. In contrast to a place like,

say, Poulton-le-Fylde in
Lancashire, its French le for plain
Yorkshire the or just t' has been
dropped.

Trencar: Near Kilburn, it means
'marsh frequented by cranes',
from ON cran + ON kjarr
'marshy land'.

Ugthorpe: Not particularly ugly.
All the name means is 'Uggi's
Village' from that Old Norse
gentleman and thorp.

Upleatham: Next to New
Marske. The Domesday Book
form means 'upper slopes',
probably from ON up hlithir.
Later spellings ending in –um
mean at those upper slopes.

Upsall: Between Thirsk and the
Hambleton Hills. From ON up-
salir 'high dwellings', which tal-
lies with its situation on a hill's
fairly steep upper slope. It is a
distinguished-looking name,
shared with Upsala, the old capital of Sweden.

Wilton: In the general area are
two such names, one five miles
east of Middlesbrough and the
other four miles east of
Pickering. There are opposing
explanations of its meaning;
'wild uncultivated land' from OE
wilde + tûn or 'willow farmstead' from OE wilig + tîn. It
does show how awkward placename research can be, but from

old spellings the second explanation seems preferable.

Yarm: Probably from OE gear 'dam, enclosure for catching fish'. Yarm is by the River Tees. Unlike modern English gear, the first sound in the OE word was y as indicated by the later Y- and J- spellings. The final m comes from its being written as a dative plural, OE gearum, meaning 'at the fish pools'.

Yearby: Near Redcar. 'Upper farm' from OE ufera + by. Named in comparison with places close by.

Yedlingham: Nine miles northeast of Malton. Another -ingas family name like Birmingham, Nottingham, etc. From Eâda + ingas + hâm. 'Home of Eâda's descendants'.

Robin Hood's Bay

EAST RIDING AND EAST COAST

Argam: (In Hunmanby). 'At the shielings, hill pastures', from the dative plural ergum of ON erg.

Aughton: North-east of Selby. 'Oak-tree farm' from OE âc + tûn.

Balkholme: East of Howden. 'Balki's field or small island' from the name of its early Scandinavian owner + ON holmr 'island'.

Barmby Moor: Near Pocklington. Despite its name Barmby Moor is a village and the moor it refers to is Spalding Moor. Barmby means 'Barns' farm'. Retention of spellings with n from Domesday Book for approaching 600 years makes this derivation likely. Perhaps the change of n to m was because both are lip or near-lip sounds and it is easy to slip from one to the other.

Barmby-on-the-Marsh: West of Howden. 'Barni's farmstead' from an Old Danish personal name + ON by. For the change of n to m before the lip sound b

compare the last item.

Bellasize: East of Howden. 'Beautiful seat' from French belle 'beautiful + assis 'seated'.

Beswick: North of Beverley. Probably 'Bessi's dairy farm' from the ON personal name + wic.

Beverley: Probably 'beaver meadow' from OE beofor + leah.

Bewick: Near Aldbrough. 'Bee farm' from OE beo + wic.

Bishop Burton: To Burton 'fortified place' from OE burh + tûn was added Bishop because the Archbishop of York held lands there. Today the Archbishop's palace is at Bishopthorpe on the Ouse a few miles south of York and is the usual turning point for boats cruising from the city.

Bishop Wilton: North of Pocklington. 'Wild, uncultivated enclosure' from OE wild + tûn. For many centuries the Archbishop of York held the land and Bishop was added to the Domesday Book name as a belated acknowledgement of the link.

Boythorpe: Near Wold Newton. 'Boia's village' from a late Old English personal name + thorp. The personal name is thought to have been from Old Danish Boie.

Brantingham: Just north of Brough. Probably 'homestead of the Brantings, dwellers on the steep slopes'. Early spellings support an origin from an OE personal name Brant varying with the spelling Brent and ing for his people.

Brantingham lies at the south-ern end of the Wolds in steep terrain and so a link with the Northern dialect word brant 'steep' has been surmised, although by the time I helped to cover Yorkshire in language studies it seemed to have died out from ordinary conversation. Times have much changed since the birth of the placename Brantingham.

Brigham: South-east of Driffield. 'Homestead by the bridge' from OE brycg + hâm. Brigham is only about half a mile from Frodingham Bridge over the beck.

Briscoe: Near Ellerby. From ON birki + skôgr 'birch wood'.

Brough: From OE burh 'stronghold'. Extensive Roman remains have been found there and the Roman road through Lincolnshire (Ermine Street) so to speak crossed the Humber to enter the East Riding by ferry at Brough and continued north on the so-called Humber Street to Stamford Bridge.

Broxa: Six miles from Scarborough on impressive moorland. 'Brocc's enclosure' from the name of its OE early owner + OE (ge)haeg.

Bubwith: East-north-east of Selby. 'Bubba's wood' from the name of a Viking + ON vithr 'wood'.

Burnby: Near Pocklington. 'Village by the stream' from ON brunnr + by.

Burton Pidsea: In Holderness. From OE burhtun 'fortified place' + Pidsea 'pool in the marsh', probably from an OE pidu 'fen' + sae 'lake, pool'.

Cavil: North-east of Howden. 'Tract of land where jackdaws are found' from an OE câ 'jack-daw (compare the sounds they make), + feld 'field'.

Cayton: South of Scarborough. 'Caega's farm' from his OE name + tûn.

Cloughton: North of Scarborough. 'Valley farm' from OE clôh 'deep valley' + tûn.

Coniston: Just north-east of

Hull. 'The king's farm' from ON konungr + by, with that ending in Old Norse by later replaced by the English equivalent tûn.

Cottam: In the Wolds south of Langtoft. Like Coutham below meaning 'at the cottages' from OE cotum in the locative case, i.e. showing where things are.

Cottingham: A Northern suburb of Hull. It was 'home of Cotta and his followers' from OE Cotta + ing (representing his people) + hâm 'home'. Nowadays instead it is very often the home of Hull University students.

Coutham: South-west of Redcar. 'At the cottages' from OE cotum.

Dalton: Apart from the necessary North and South to distinguish them, North and South Dalton, five miles from each other on the Wolds south-east of Driffield, have the same word developments, meaning for each 'valley farm' from OE dael + tûn. Nothing surprising there.

Drax: 'Long stretch of river'. A 1208 spelling has Langrak. From OE lang 'long' + OE hypothetical racu.

Driffield: Is according to its name history a 'dry field' from OE dryge + feld.

Now please ignore the next minor anecdote, if, as is almost certain, you have far more significant travel experiences. It is inserted as a change from lists and because it did concern Driffield. As a fairly new motorist in thick mist I had just, as I later found, reached a hilltop outside the town when my car died. Then it awoke, barely crawling at first with no engine help but then faster and ever faster till after half a mile it lurched for me into a garage where brakes were desperately slammed on. To my surprise and delight, a mechanic immediately reconnected a hosepipe, no charge, and away we went. Driffield may be in word history a dry place, but could apparently produce a liquid, petrol, from nothing.

Dringhoe: Near Skipsea. The first element is dreng, commonly used for a certain type of free tenant, whilst the second element has been interpreted as from OE hoh 'height, hill', though the Domesday Book spelling Dringolme suggests it could be OE holm 'water meadow' with the m sound vanishing through influence of hoh. Slightly puzzling.

Duffield: In North Duffield and South Duffield, both east of Selby, Duffield means 'tract of land frequented by doves' from an OE dûfe + feld 'field'.

Easington: On the far tip of Holderness. 'Farm of the descendants of Ësa' like the North Yorkshire place with a similar name.

Eastburn: South-west of Duffield on Eastburn Beck, and meaning 'east spring or stream'. It is hard to say whether the two words making the name are old English or Scandinavian, i.e. whether they are OE eâst + burna 'stream' or their ON equivalents austr and brunnr, though the meaning is not affected.

Ellerker: West of Hull. 'Alder-tree marsh' from OE alor + ON kjarr.

Ellerton: North-east of Selby. 'Farm of the alders' from OE alor + tûn like the Ellerton near Catterick.

Elloughton: North of Brough. Site of a Scandinavian pagan temple, so understandably 'farm or hill with a heathen temple' from OE tûn 'farm' or dun 'hill' + ON helgr 'temple'.

Everingham: West of Market Weighton. 'Homestead of Eofor and his people'.

Fangfoss: East of York. Probably 'Fangulf's ditch (Latin fossa)'.

Fell Briggs: Near Marske. 'Plank bridges' from either ON fjol or, with th becoming f, OE thel,

both meaning 'plank'; ON bryg-gya 'bridge'. Connected too is OE fellan 'to cut down trees'.

Filey: Probably a hybrid, 'cot-ton-grass glade', from ON fîfa or fifill (the grass) + OE leah.

Filey Brigg is a narrow ridge of rock extending half a mile into the sea. Its second word, unlike bridge though of the same meaning, must come from ON bryggya because of its g sound. Though Stourton just south of Leeds was the traditional area, Filey was the site of the best thick rhubarb I have seen – in a small garden by the railway crossing. Filey may be famous for its fishermen's choir, but it must also nurture some splendid gardeners.

Flamborough: 'Fort of a Norseman called Fleinn'. ON fleinn also meant a hook so the name given to Fleinn might also have meant 'sharp-tongued'. Danes' Dyke at Flamborough Head was not dug by the Danes but was there long before they came.

Folkton: 'Folki's farmstead' from ON Folki + tûn.

Fitling: South of Aldbrough. 'Settlement of Fitela and his people', from Fitela + OE ingas 'followers'.

Flixton: Near Filey. 'Flik's farm' from the ON personal name +

tûn.

Flotmanby: On the chalk escarpment near Filey. 'Site of Vikings'. Flotman was an Anglo-Saxon word for a Viking. It was coined by neighbours, not the Vikings themselves, probably from a nickname from ON flot 'ship' and mathr 'sailor'.

Fordon: Near Hunmanby. It means 'in front of the hill' from OE fore + dun. The topography of the district agrees as Fordon lies below a steep hill hemmed in by valley arms on either side.

Foston-on-the-Wolds: Between Driffield and Skipsea. 'Fot's farmstead' from ON Fôtr + tûn. Foston (see York section) has the same word development.

Frodingham: 'Settlement of Frôda's people', from the OE personal name + ingas 'follow-ers'. North Frodingham is safely established on maps, but South Frodingham near Withernsea about 20 miles away from its 'twin' seems to have disappeared from most of them.

Full Sutton: East of York like various Fulfords in England has a name more awkward than it seems. Sutton from OE sûth + tûn 'south settlement' is harm-less enough; but Full (OE fûl) means 'dirty', perhaps used contemptuously.

When with a colleague I was kindly allowed after strict vet-ting to visit a prisoner there because of an appalling crime to see whether help was possi-ble or indeed reasonable, out-side York railway station my col-league remarked how sheepish-looking were relatives and friends waiting for the bus to take us to the prison, but that seemed a great pity because the crimes were not theirs.

Ganton: On the Wolds Way eight miles from Filey. 'Galma's farmstead' from that personal name + tûn.

Goole: Previously in Lincolnshire, now considered East Riding. ME gool 'small stream, ditch, sluice', first recorded 1552.

Goodmanham: A village next to Market Weighton. 'Home of Godmund and his people'. It is of the –ingham type of place-name (compare Sandringham, Immingham, etc.) meaning 'home of the descendants/fol-lowers of …' Long ago Bede referred to it as Godmundingham and in the 7th century it was sufficiently important to be the site of the royal residence for the ancient kingdom of Deira.

Hackness: At the foot of a very prominent headland behind

Scarborough. Stanley Ellis in a most useful article for the 1990 Yorkshire Dialect Society Transactions states 'The Venerable Bede mentions Hackenos 'headland of Hacce'. –Ness sounds Scandinavian but cannot be as Bede preceded Scandinavian invasions. It will be for –nos.'

Haltemprice: Name given to the priory of Augustinian canons which moved to Cottingham near Hull in 1325-6 and shortly afterwards to a place nearby, now lost, called Newton. The name meant 'great enterprise' from French haute emprise. The interesting point for an enthusiast of place-names is that here a French name replaced an English one just as occurred with Mount Grace Priory near Osmotherley and on the Cleveland Hills.

Hanging Grimston: Near Painsthorpe on the Wolds. Grimston equals 'Grim's farm-stead' from ON Grimr + tûn, whilst Hanging from ON kengja 'to hang' describes its situation, 'overhanging, sloping'.

Harswell: Near Market Weighton. 'Hersa's well or spring' from OE Her(e)sa + w(i)ella.

Harwood Dale: North-west of Scarborough. ON har 'rock' would agree with the natural surrounds, but more likely it is an animal-inspired name from OE hara 'hare' + wudu 'wood' + dael 'dale'.

Hawsker: On the cliffs south of Whitby. 'Hawk's enclosure' from the ON personal name Haukr + garth. The Hawsker Bell is a nearby foghorn to warn ships of dangerous rocks along the shore.

Hayton: Near Pocklington. 'Hay farm', OE haeg + tûn, farms being often named from their crops.

Hedon: East of Hull. Probably OE haeth 'uncultivated land' + dun 'hill'.

Heighholme: Near Hornsea. 'Water meadow used for hay'. Another place named from what was grown there, from OE (ge) haeg + ON holmr 'small island'.

Hesselskew: Near Market Weighton. 'Hazel wood', a mixture of OE haesel + ON skôgr 'wood'.

Hilderthorpe: At the southern edge of Bridlington. 'Hildiger's village' from ON Hildiger + thorp with g lost between the sounds d and th.

Hinderwell: Near Staithes. 'Hild's well'. From Saint Hild, whose monastery was a few miles away, made to resemble Hildr, the same name in ON, +

OE w(i)ella.

Hive: East of Howden. From OE hyth 'landing place on a river', though Hive is no longer on a river or navigable drain. Dialect changes of voiced th to v are not uncommon.

Holme-on-the-Wolds: North East of Market Weighton. From ON dative plural i haugum 'on the hills' or OE hôhum with similar meaning 'on spurs of land or hills'.

Hornsea: 'Lake lying in a projecting piece of land' from OE or ON horn 'corner of land' + OE sâe 'pool, lake'. Hornsea was originally, as befitted its name, the name of the great lake called Hornsea Mere.

Howden: From OE heafod 'head' + OE denu 'valley'.

Hull: The earliest name of this settlement was in 1160-80 Wyk, probably from ON vic 'creek, inlet' referring to dwellings at the mouth of the River Hull a tributary of the Humber. In 1292 Edward I exchanged lands with monks to found what was officially called, in Latin as well as English, Kingston-upon-Hull. In 1382 the mayor and townsfolk were granted a port below the town to be called Hull. After the 15th century, especially in documents where the name did not have to be repeated many times, the longer name Kingston-upon-Hull was often employed; but the shorter and convenient name Hull has persisted in most official and certainly popular use until now. Hull is a great fishing port and should not be maligned as part of the saying "From Hull, hell and Halifax oh Lord deliver us". I have experienced two of the group without ill effect.

Hunmanby: 'Dwelling of dog-keepers or houndsmen', from ON hundemanna + by.

Hutton Bushell: South-west of Scarborough. 'High abode (OE hôh + tûn) of the Bushell family', a partly French name for a French family which owned the land in th 12th and 13th centuries.

Irton: Near Scarborough. 'The Irishman's or Irishmen's farm' from ON Iri, used of a Scandinavian who had been in Ireland, + tûn.

Keyingham: In Holderness. 'Homestead of Câega and his people' from OE Câega + ing + ham.

Killerby: South of Scarborough. 'Ketill's abode' from an ON personal name Ketill or similar + ON by.

Kilnsea: Probably 'kiln by the pool' from OE cyln + sâe 'sea, lake, pool'.

Kirby Underdale: Halfway between York and Driffield. 'Church of Hundle's valley', from kirkja 'church' + by 'settlement' + personal name Hundôlf + daelr 'valley' – four Scandinavian placename parts.

Kirkburn: Near Driffield. The first spellings like Domesday Book's Burnous 'house by the stream' suggest an OE origin of burna + hûs.

Knedlington: Near Howden. 'Farmstead of Cneddel and his people'. The hypothetical OE personal name Cne(o)ddel would be a pleasant name meaning 'eager, diligent'.

Lythe: From ON hlith 'slope', referring to one near the coast north-west of Whitby.

Mappleton: South of Hornsea. 'Farm by a maple tree' from OE mapel + tûn. Many placenames have for their first elements names of trees.

Market Weighton: For its second word various suggestions have been aired in print. It might have been named from some man such as Weah, Wigheah or Wiht; or, as it lies on a Roman road, Latin vicus 'settlement' should be considered. The Market element of the name is easier to explain. It is a fairly late addition to distinguish it from Little Weighton, though the latter is eight miles away and Market Weighton was granted its charter as long ago as 1252.

Marske-by-the-Sea: 'Marsh(es)' from OE mersc. After Domesday Book the spellings show substitution of the Scandinavian sounds sk for the old English sound sh spelt sc.

Nafferton: Near Driffield. 'Natfari's dwelling' from an apparent ON nickname Nâttfari 'night traveller' + tûn. What the night-traveller did at night is not revealed. It could well be a name coined by friends or neighbours who liked to poke fun.

Osgodby: There is an Osgodby south-east of Scarborough and another north-east of Selby. Both are derived from an ON personal name Âsgautr and mean 'Asgaut's farm'. The name occurs also in mysterious Danish runes.

Oubrough: Near Ellerby. 'Owl-haunted stronghold' from OE ûle + burh.

Outgang: A common word for the way out from, say, a large field as e.g. around Thorpe Bassett. From OE ûtgang or ON ûtgangr.

Owsthorpe: Near Sandholme.

'East village' from ON austr + thorp.

Owstwick: In Holderness. 'East dairy farm' from ON austr + wic.

Painsthorpe: Halfway between York and Driffield. 'Pain's hamlet' from French personal name Pain from Pagan(us) + thorp (here for his outlying hamlet of Kirby Underdale).

Patrington: In Holderness. First element doubtful, though it may be associated with St Patrick + tûn 'settlement', which would give 'abode of the followers of St Patrick'.

The amazing language feature to me, dialect-searching in the 1950s in Patrington and other parts of Holderness, was the sporadic but notable absence of many a 'the'. Folk would say e.g. not "Put it into the cart" or even "put it into t'cart" with a kind of bitten-off t called a glottal stop that many true Yorkshiremen employ, but simply "put it into cart". It might seem a small matter but my 'boss', Professor Orton, who incidentally hated being known by that title on any dialect expedition, believed it very historical, springing from a lack of 'the' in Old English. Whatever the reason, it was then certainly missing in Holderness. According to the literature, Holderness is the only part of England where this absence has been recorded, and it would be very interesting to know how much, if at all, it is still lost.

Paull: A fishing village east of Hull. It has been explained as a contraction, 'nook of Pagela, a spelling for it in Domesday Book, + OE ealh 'corner'; but an OE supposed pagol 'stake or landmark to guide ships' is not impossible as Paull lies on the bank of the Humber. Paull, as found on our searches along the coasts of England and Wales, harbours many fishing terms strange to a landlubber such as cat's paw 'rough water', truck-ships, 'small fishing-smacks', web 'oar-blade' and loom 'oar-handle'.

Pickering and the North York Moors Railway: Pickering's name comes from the OE Piceringas and means 'settlement of Picer and his people'. Pickering Castle like so many others is in ruins. In the 16th century wagonloads of stone and slate were taken from it to improve Sir Richard Cholmey's own home, now gone, near Thornton Dale. Pickering is now best known as the start of the North York Moors steam railway, lovingly kept in operation by enthusiasts and meandering 18 miles through unspoilt scenery to Grosmont, very different from the 1830s when George Stephenson had to organise a track for it over a bog in Newtondale.

Potter Brompton: The origin of Potter is doubtful but it may be an occupation name. Brompton is from OE brôm + tûn, meaning 'enclosure overgrown with broom, gorse'.

Potto: South-west of Stokesley. 'Hill near the small valley'. Compare old Cumbrian dialect pot 'deep hole' and OE hoh 'mound'.

Raisthorpe: On the Wolds. 'Hreitharr's village' from ON Hreitharr + thorp.

Ravenseer Odd: Ravenseer means 'Hrafn's sandbank' from the ON personal name Hrafn + eyrr. The suffix Odd is very suitable, for it was a port at the mouth of the Humber with a very odd history. It stood a little south-west of Kilnsea near Spurn Point. Salzman in his English Trade in the Middle Ages states that the sea piled up sandbanks on which grew a port and nest of pirates menacing the prosperity of Grimsby. It became quite well-known as a trading port and, though this needs confirmation, for a time even returned its own member to Parliament; but sadly by

about 1360 the very seas which had created it washed it away.

Rawcliffe: Near Goole, having the same word development as that for the Rawcliffe close to York. It means 'red cliff' from ON rauthr + ON klif.

Ridgmont: Near Burstwick in Holderness. 'Red hill'. Judging by its old spellings, it may have been named from Rougemont in France, which is from Old French rouge + munt. Very few English placenames come from France.

Rise: South-west of Hornsea, meaning 'among the brushwood'. From OE hrîs in a dative plural form hrîsum. Perhaps a place where faggots were collected.

Robin Hood's Bay: Name of both bay and village. The name was not found before 1352 and probably came from popular ballads. The Yorkshire name presumably, like that of the outlaw, is fictitious. He could never have visited this picturesque place, for its name does not crop up until three centuries after Robin's supposed death (see e.g. article by Stanley Ellis in The Yorkshire Dialect Society's Transactions for 1990).

Rolston: Near Hornsea. 'Rolf's farmstead' from an ON personal name Hrôlfr + OE tûn.

Rhos: A hamlet north-west of Withernsea. From Celtic rhos 'moor, heath'. A Domesday Book spelling Rosse shows what the vegetation was like in those distant times.

Rudston: East of Bridlington, meaning 'the rood, i.e. cross, stone' from OE rôd + stân. A place associated with Christianity, named from a great stone pillar in the churchyard. There was an OE word rudu 'redness' and that can come from weathering or iron in sandstone, but the underlying element in the name Rudston is the cross. Compare *The Dream of the Rood*, a tender Anglo-Saxon poem where Christ's cross itself sheds tears because of what is happening on it.

Runswick: Perhaps 'Raegan's creek' from OE Raegan or ON Hreinn + ON vik.

Ruswarp: 'Silt land overgrown with rushes', from OE ryse 'rush' + warp 'river sediment'. In December 1914 early in the Great War the German navy bombarded Ruswarp along with Whitby and a coastguard in a cliff top post was killed. Now it is more peaceful; my honeymoon in Ruswarp confirmed the quiet and excellence of walks from it over the moorland.

Ryhill: South-west of Hedon. This is another placename denoting a local crop for it means 'rye-hill' from OE ryge + hyill.

Saltburn-by-the-Sea: 'Salt stream' is the basic meaning from OE s(e)alt + OE burna, probably because of the mineral salt found in this district. In Victorian times it was a new seaside resort. Everyone's recollection is different, mine being of a delightful Indian summer there operating from a boarding-house headquarters, but most people find Saltburn even in less ideal weather quite agreeable.

Sancton: Near Market Weighton. 'Sand farm' from OE sand + tûn, with sanct spellings through popular association with OE and ME sanct, ME saint. It belongs to a large group of placenames describing the locality.

Scalby: One Scalby is next to Scarborough and another miles away east of Howden, but both have the same meaning and etymology, namely 'Skalli's farm' from the name of that rather ubiquitous Scandinavian gentleman.

Scampston: Near Malton. 'Skammr's dwelling' from an ON nickname + tûn.

Scarborough: 'Skarthi's

stronghold', from Skarthi, a Norwegian apparently so named for his harelip, + ON borg.

For a long time it has had something of a military history (see the Places of Interest section) and during the last war we were trained to drive vehicles through the water up Scarborough beach, as became evident later to prepare for the Normandy landings. Scarborough will not like being compared with Blackpool – it prides itself on being a little more refined – but both have been and are very friendly places.

Seamer: Near Scarborough. It probably means a marshy lake, now drained, from OE sâe 'sea, lake' + OE mere 'pool'. Another Seamer near Yarm has developed word wise in a similar way.

Settrington: Near Malton. Meaning doubtful but most likely from OE sâetere 'robber' or similar-sounding personal name + in + tûn, to mean probably 'place of the robber and his gang'.

Sherburn: Under the Wolds. 'Bright, clear stream' from OE scir + burna.

Silpho: Near Scarborough in excellent moorland. Probably 'Sylve's hill' from an Old Danish personal name Sylve + ON

haugr 'hill'.

Skeffling: In Holderness. Hard to decide whether its name is English or Scandinavian, but perhaps it arises from an OE personal name Sceftel + ingas to mean 'abode of Sceftel's people'.

Skipsea: 'Lake where a ship can sail' from ON skip 'ship' + ON sâer 'sea, lake'.

Skipwith: From Old Northumbrian Scîp(a)wic 'sheep farm', with ON substitution of the sounds sk for OE sh (spelt sc) and ON vithr 'wood' for OE wic.

Sleights: 'Flat, level grounds' from ON sletta. According to tradition (see Rhea, *Portrait of the North Yorkshire Moors*), the chapel 1.5 miles upstream is where a scene in Sir Walter Scott's *Marmion* is set, whilst at Sleights Lane End there crashed on 3rd February 1940 a Heinkel bomber, the first enemy bomber shot down in England in World War 2. The successful pilot was Group Captain Peter Townsend, former close friend of Princess Margaret.

Sneaton: Near Whitby. 'Snjôs farm' from Old Danish Snjô + tûn.

Southburn: Near Great Driffield. 'South stream'. There is some doubt whether the name stems from OE or ON but the OE elements are recorded a little earlier. Thus it is probably from OE sûth + OE burna.

South Cave: No sign of a cave for sheltering. The name is apparently from OE câf 'swift, quick', applied first to the stream which flows rapidly down the Wolds through North Cave and near South Cave, a village which then grabbed the stream's name, rather resembling how Hull acquired its name from its own River Hull.

Spaldington: North of Howden. 'Farmstead or village of people of the Spalde tribe', from Spaldinghas, the OE name for that large tribe established in Lincolnshire, + tûn.

Staithes: A little fishing village built in a creek on the coast. Early spellings suggest OE staeth 'landing place', not the equivalent Old Norse word.

A Leeds research student, before marriage Mabel Lawson, who courageously went out to sea with the fishermen, concluded that when at sea they spoke another language altogether, but more probably it was just their special variety of English. It would have to include their terms like kessen bowl 'inflated dogskin float' and wither 'barb on a hook' not met by helpers and me in later fishing-word searches round England and Wales and others like mop 'cod about 1-2lb'. and clog 'oar-handle' which were. Only Staithes fishermen know whether they speak part of a foreign language out at sea.

Stamford Bridge: 'Bridge by the stone-paved ford' from OE stân + ford + brycg. Site of King Harold's 1066 battle against the Danes. We hear far more of the battle forced on him that year on a 'second front' at the Battle of Hastings, about which even now experts are not all agreed how he died – certainly guarded by his closest retainers and defending from a mound, but shot by an arrow or (despite the Bayeux tapestry) hacked to death?

Stillingfleet: North of Selby. 'Stretch of the Ouse belonging to Styfela and his people', from Styfela + ing + OE fleôt 'river'. Styfela is also an ingredient in another of our Yorkshire place-names (see Stillington in the York Section).

Stockholm: Near Thorngumbald in Holderness. 'Island cleared of trees' from ON stokkr 'tree stump' + ON holmr 'small island'. East Riding Stockholm is named like and probably has the same general word history as Sweden's but

hardly of course the magnifi-
cence of that capital.

Summergangs: Recorded from
the Hull and Holderness areas.
'Paths which can be used only
in summer'. From OE sumor or
ON sumar + OE or ON gang
'road, track'.

Thickleby or Thirtleby: (Spelt
differently in reference books –
take your pick.) Near Burton
Constable in Holderness.
'Thorkell's farmstead' from the
ON personal name + ON by. A
Thirkleberry with the same word
development is found elsewhere
near Kilburn (see the Moors section).

Thorpe: North-west of Beverley,
meaning simply 'farm' from OE
thorp. It is unusual to have this
standing alone on a map, but
spellings show that from 1200
was added the feudal owner's
name, probably Winemund.

Thorpe Bassett: Near
Rillington. From OE thorp 'outlying village, hamlet' andname of
a previous owner, William
Bassett, who held the land there
in 1204. This usage, with a
word before or behind thorpe
to show which thorpe is meant,
is far more usual.

Thorpe-le-Street: Near Market
Weighton. 'Village on the road'
from OE thorp + French le 'the'
+ OE strâet, the last word refer-

ring to the Roman road on
which the village stands.

Towthorpe: On the Wolds
south-east of Malton. 'Tove's village' from ON Tôfi + thorp. A
similarly named village is near
York (see that section). It is con-
fusing to have so many villages
of like names, but people seem
reluctant to change them.
Besides in fairness which should
stay and which would have to
change its name?

Ugglebarnby: Near Whitby.
'Owl-beard's farm' from an ON
nickname Uglubarthi + by,
though it is unclear what exact
type an owl-beard was. The
change in the placename's ending is by association with another nearby called Barnby just
across the River Esk.

Uncleby: 'Hunkel's farm' from
ON Hûnkell + by.

Warter: On the Wolds between
Pocklington and North Dalton it
signifies 'gallows-tree' from OE
wearg 'felon' + OE treôw, a
social custom no longer
demanded.

Wauldby: East of Hull. 'Farm
on the Wold' from OE weald +
ON by.

Weel: East of Beverley. 'Deep
place in the river' from OE wâel.
Weel is on the River Hull.

Weeton: In Holderness just like
Weeton north of Leeds (see

York section); 'willow farm' from OE withig + tûn.

Wharram-le-Street: South-east of Malton. 'At the bends' from ON hvarfum 'bends' + French le 'the' + OE strâet because it stands on an ancient Roman road. It is a deserted medieval village around which passes The Wolds Way.

Whitby: Its first name was OE Streonshalh 'Streon's nook of land', but by the time of Domesday Book it had become Witebi etc., 'Hvîti's farmstead' from the ON by-name Hvîti and ON by. It was an old whaling port and is a thriving fishing port and sailing centre. Its surroundings, pleasantly named and less so, include **Ling Hill** from ON lyng 'heather' and **Murk Head** 'Dark Hill' from ON myrkr + OE heaford. For other details of it, see the Places of Interest section.

Willerby: Places so named are near Scarborough and near Hull. They mean 'Willard's farm' from OE Wilheard.

Withernsea: Perhaps 'pool near the thorn tree' from OE with or ON vith 'beside, near' + thorn 'thorn tree' + OE sâe 'sea, lake, pool'. The pool or lake has disappeared.

Willitoft: North of Howden. 'Homestead near the willows' from OE wilig + ON toft.

Wilsthorpe: Near Bridlington. 'Wifel's village' from ON Vifill + thorpe.

Wintringham: 'Home of Winter's people' from OE Winter + ingas 'descendants' + hâm 'home'.

Wykeham: Near Scarborough. 'Small settlement near a Roman Station' from OE wîchâm.

Jon A. 14S '97

Middleham

OTHER PLACES OF INTEREST

Beningbrough Hall: Eight miles north-west of York. Beningbrough means 'Benna's fortified place' from his personal name + OE burh. National Trust property. Historic house and gardens.

Bolton Castle: In Wensleydale. From OE bothltûn 'settlement with a special building' + castel. Where Mary Queen of Scots was imprisoned. In 1379 Richard le Scrope built a castle there. In the 17th Century, when Wensley, which gave its name to Wensleydale, was failing to attract much trade as a market centre, a man later to become Duke of Bolton instead developed this castle four miles to the west and prominently high on a steep slope.

Bolton Priory: From OE both-ltûn as above + French priorie. It is near the River Wharfe in a picturesque setting six miles from Skipton-in-Craven.

Burton Agnes Hall: A Norman manor house between Driffield and Bridlington, the ruins of which are carefully preserved by the National Trust. Burton Agnes itself is an attractive village. Its name comes from OE burh + tûn and means 'fortified farmstead or manor house' and to it in the 13th century was added the name of Agnes, who belonged to the important landowning family mentioned in Shakespeare. Agnes de Percy had been associated with the village around 1150. In those days quite a number of women held estates and dealt with country matters in a status roughly equal to men's.

Burton Constable Hall: Historic house near Hull. From Burton (see preceding item) and Constable, a family name. In the 12th century the widow of Gilbert de Alost married into that family, so that Constable was added to the placename and from 1294 onwards associated documents often refer to that family.

Byland Abbey: Three miles

south of Kilburn's White Horse. Byland comes from an OE personal name Be(a)ga and means 'Beaga's land'. The best way to appreciate it is to wander undisturbed around its vast empty ruins.

Captain Cook's Monument: It stands just south of Roseberry Topping on the edge of the moors near Great Ayton ('farm by the river', OE â + tûn). Before he worked in a Staithes grocer's and later went to sea to become a famous explorer. Captain Cook spent most of his early life in this area, like his father working for a time on a farm on the southern slopes of Roseberry Topping, perhaps just as fitting an area for him to have a monument as another to him prominently overlooking Whitby harbour, which sadly explains that he was eaten by cannibals.

Castle Howard: Historic house six miles west of Malton in the aptly named Howardian Hills. From a York to Malton bus just a glimpse of it can be caught, but tourists will need to go much nearer. Howard appears in the name because it is a modern mansion built by the Howard family. It is certainly Yorkshire's finest private residence, built for Charles Howard, Third Earl of Carlisle, whose descendants still live there. Its site is dramatic, between two lakes with extensive gardens and glorious fountains. The older name for a settlement very close to it was Hinderskelf, from ON Hildr, a woman's name, + ON skjâlf 'small plateau', which can be seen just to the west.

Jervaulx Abbey

Druids' Temple: In a plantation a little east of the reservoir at Leighton ('glade hamlet' from OE leâh + tûn) this fine folly of a massive druids' temple was built around 1800 by William Danby of nearby Swinton. It is worth a visit to wonder at the sheer enormity of it.

Duncombe Park: Near Hemsley (probably OE dun 'hill' + certainly cumb 'valley'), a baroque mansion amid beautiful gardens near the River Rye, ancient trees and a ruined Norman castle with formidable earthworks.

Eden Camp: Just outside Malton. Presumably named from Eden House on the site. It has 14 or so huts with relics of military and domestic items from the Second World War. Sometimes a brass band plays. Worth a visit by veterans of those awkward times and their descendants who know nothing about articles then in use like milk ladles, coal scuttles, toasting-forks, etc.

Guisborough Priory: Guisborough is from a rare ON name Gîgr + OE burh, resulting in Domesday Book names like Ghigesbure. Best of the remains is the ruined east (altar end) window.

Hallikeld Hall and Spring: North-east of Northallerton.

Hallikeld means 'holy spring' from a hybrid mixed-race name OE hâlig + ON kelda. Originally hâlig had a pagan connotation but no English placename is known where this is definitely so.

Harewood House: Seven miles from Leeds or only 15 minutes by car outside the rush hour from Leeds or Harrogate. Harewood's word development is puzzling. Its first element might be OE hara 'hare', har 'rock' or hâra 'grey', though the second element is almost undoubtedly OE wudu 'wood'. It is the home of the Queen's cousin, the Earl of Harewood. It contains priceless treasurers, royal souvenirs, magnificent Adam interiors, Chippendale furniture, etc. and most visitors are awed by the lakeside aviary with well over 100 rare species.

Hovingham Hall: Hovingham, north of the Howardian Hills and Castle Howard, and meaning 'home of Hof's people', is from OE hof, originally 'temple' but used as a personal name, + ingas 'descendants' + hâm 'home'. A few years ago The Yorkshire Dialect Society were most honoured when their president, Sir Marcus Worsley, invited them to his home, Hovingham Hall, for one of their meetings. It has a private cricket ground, among the oldest in England. What do Arundel and the Hambledon Club have to say to that?

Jervaulx Abbey: On the River Ure four miles south-east of Leyburn. Jervaulx means 'The Ure valley', from the river name + OFr vals. It was the home of a Norman monastic foundation.

Kirkham Priory: North-east of York. Kirkham stands for 'village with a church' from ON kirkja + OE hâm. The Augustinian priory lies at the start of a picturesque gorge cut by the River Derwent. Its ruins include a wonderful carved gatehouse.

Middleham Castle: Middleham needs little explanation, meaning 'middle settlement' from OE middel + hâm. Middleham Castle governed the entrance to Wensleydale. It is only one of many such castles scattered round Yorkshire, particularly in the North Riding. It was from Middleton Castle in 1483 that Richard, Duke of York, rode south to become King Richard III.

Mount Grace Priory: Was a Norman-French building which quite submerged an English one. Motorists on the A19 near Osmotherley can see its ruins near the moorland slopes above. In Domesday Book its name was not French but of mixed race, 'Bordel's farm' from

the Anglian personal name Bordel + ON by. The change to French Monte Grace did not show in records till 1413 though the Bordelbi type persisted till 1508, showing how late came any French influence on English placenames.

Ormesby Hall: National Trust property in south-east Middlesbrough. Ormesby 'Orm's farm' as from the ON personal name + by.

Parcevall Hall and Gardens: Near Appletreewick. The building is an Elizabethan mansion supposed to have hidden notorious highwayman William Nevison. For Parcevall 'Percival' compare what has happened in spelling and pronunciation to our words like Derby, sergeant and clerk.

Rievaulx Abbey: Near Helmsley. Preserved by the National Trust. Rievaulx means 'Rye valley' from the River Rye, which flows past it, and OFr vals. Unlike, say, Mount Grace Priory, where the OE name was completely obliterated by a French one, this abbey founded by French monks was given a French one from the start. The monks who established it in 1132 came from a Cistercian order in Burgundy putting much emphasis on scholarship and ritual but the monks breaking away to North Yorkshire believed more in austerity and toil.

Sledmere House: In the High Wolds. Sledmere means 'pool in the valley' from OE slaed 'valley' and mere 'pool'. There are several pools nearby, but that causing the name is probably the one on the hillside near the church. This historic house is very popular with visitors.

White Horse of Kilburn: Nestling under the Hambleton Hills. Kilburn equals 'Cylla's stream' from OE Cylle + OE burna or ON brunnr. Helpers find some difficulty grooming the Kilburn White Horse just north of the village as it is not cut from chalk like others in the South of England but from a dull brownish limestone and needs regular touching up with white materials. Over 20 people can sit on its eye and it can be seen from afar, a tribute to whoever decided to leave this memorable giant mark on the moors.

Hutton-le-Hole

RIVERS AND WATER

Yorkshire unlike the Lake District does not abound in lakes, but we do have west of Aysgarth in Wensleydale a natural lake, lovely **Semerwater**. In Domesday Book this was written Semaer from sae 'lake' + mere 'pool', the latter being added to reinforce sae in the sense of 'lake' rather than its other meaning 'sea', as the former meaning was dropping out of common use. It is a glacial lake, remnant of a bigger one, and still gradually shrinking. A small lake high among mountains or on a moor is a tarn from ON tjorn. Yorkshire's best-known representative is **Malham Tarn**, noted for its trout.

Frequent are the Yorkshire becks, moorland streams named from ON bekkr. For example, **Blean Beck** from Scandinavian blaeing bekkr 'dark stream' or **Eller Beck** on Goathland Moor which after three miles joins Wheeldale Beck. People tend to think becks unimportant but some are not, such as **Pickering Beck** running 12 miles from Goathland Moor into Newtondale, and **Hodge Beck** extending from high on Urra Moor down towards Kirkbymoorside.

Waterfalls too can be spotted. The most often visited of these waterfalls or forces (from ON fors, later foss) are the falls at Aysgarth. Others are **Hardraw Force**, a cataract near Hawes, and near Ingleton, with its series of rapids and cascades, **Thornton Force** (OE thorn + OE tûn + ON fors).

For crossing a river a word more prominent in parts of the North than ford is wath from ON vath. So, for instance, we find on Bransdale Moor **Slape Wath** 'slippery ford' from ON sleipr

+ vath, a **Wath** near Cringle Moor, **Cow Wath**, **Hob Hole Wath**, **Leeming Wath** near Bedale, and many others. Then there is the **Strid** (from OE strîdan), a dangerous striding or jumping place across the River Wharfe near Bolton Abbey. The water below is deep and the legend goes if anyone falls back on the sloping stones and tries a second time, he will not be seen again.

Many Yorkshire river names are connected with Celtic. This is hardly surprising since, as pointed out by R. L. Thompson, (*Yorkshire Dialect Society Transactions for 1964*), as raiders penetrated new territory, they would best remember the existing names of rivers giving them access to it.Here are the main North and East Riding rivers:

River Aire: Probably from Old Celtic Isara 'strong river' (according among others to Eilert Ekwall, an expert famous in placename studies).

River Alne: Compare Welsh gwyn 'white' (Ekwall).

River Bain: Probably from ON beinn 'straight' (See Stanley Ellis, Yorkshire Dialect Society Transactions,1973).

River Cover: 'Hollow, i.e. deep brook' from Welsh cau 'brook' + gofer 'a rill'.

River Derwent: Long tributary of the River Ouse, flowing from The North Riding Forest Park to meet the Ouse at Barmby-on-the-Moor. Presumably Celtic dervâ 'river where ash trees were common'.

River Dove: Flowing from the North York Moors down past Kirkbymoorside. Holder of a famous rivername from Celtic du wy 'black water'.

River Esk: A good salmon river flowing into Whitby. A British river name Escâ, probably meaning just 'river, water' and a name identical with Devonshire's River Exe.

River Greta: Joining the River Tees. From ON griôta 'strong stream'.

River Humber: Probably connected with British hu-, hy- 'good, well' and ambre 'river'.

River Leven: Tributary of the River Tees, meeting it at Stockton. May well be from Welsh llyfn 'smooth'.

Murk or Mirk Esk: Starts near Beck Hole and meets the River

Esk at Grosmont. So named because, as it descends from the moors, its colour is sometimes deep brown.

River Nidd: Probably Celtic from a word root nei 'to be brilliant, shining'.

River Ouse: A common English river name: books suggest there are at least eight River Ouses in the land. From OE Ûse 'water' connected with Usa in Scandinavian sagas.

River Ribble: May be from a projected OE rîpel 'tearing' from OE rîpan 'to reap' originally 'to tear'.

River Riccal: Running south from Helmsley Moor. A tributary of the River Rye, it was called by Ekwall 'calf of the Rye, little Rye'. Calf (OE cealf) seems a likely derivation, though the mother-child comparison should not be pressed too far, since a mother feeds her baby whereas the baby-calf Riccal feeds its mother.

River Rye: Rising just east of the Hambleton Hills, it joins the River Derwent near Malton. Perhaps connected with Latin rivus 'stream'.

River Seph: Flowing gently down Bilsdale to meet the River Rye. 'Calm stream'. Compare Old Swedish saever 'calm, slow'.

River Seven: In Rosedale. Linked in name with the River Somme of Great War notoriety and even with Sanskrit sôma 'beveridge, etc.'

River Skirfare: Running through Litton, Arncliffe and Hawkswick to meet the River Wharfe. 'Bright stream' from ON skîrr 'bright, clear' + a word for 'brook' from ON fara 'to go'.

River Swale: From OE swelh 'whirlpool'.

River Tees: Related to Welsh tes 'heat, sunshine'. It may mean 'boiling, surging river'.

River Ure: Wensleydale tributary of the River Ouse. From British Isura, the old name of Aldborough, a Roman fort near the river, which eventually resulted in the ME name Yôr with a pronunciation which dialectally has almost died out.

River Wharfe: 'Winding river', connected with ON hverfr 'winding'.

River Wiske: Flowing near Northallerton. From OE wisc 'damp meadow', from which could come the meanings 'meadow stream' and then just 'stream'.

Darnholme

MOUNTAINS, HILLS AND DALES

Airedale: Named from the River Aire, probably from Old Celtic Isara 'strong' river' + either OE dael or ON daelr, though it is often hard to tell which is in such a name. Because of this doubt, such valley names will from now on usually be explained simply by the current word dale.

Arkengarthdale: Tributary valley in Swaledale. 'Valley of Arkill's enclosure' from ON personal name Arkill + ON garth + dale, this last element dale not being added till the time of Elizabeth I. Lead-mining once stirred activity in this isolated dale.

Bedale: Probably 'Beda's piece of land' from OE personal name Beda + dale.

Beedale: Extending from the North Riding Forest Park to Wykeham. It means 'booth valley' probably from OE bôth + ON dalr with a change from oh to ee-er type vowels as has been common in North Yorkshire. Beedale, pronounced like the last item Bedale, though definitely shorter on the ground has the longer spelling.

Bishopdale: Joins Wensleydale at Aysgarth. Probably 'Bishop's valley' from an OE personal name Bisceop + dale.

Black Hambleton: (399metres) From which the Hambleton Hills take their name. From OE hamel 'maimed in the sense of scarred or injured' + dun 'hill'. Most of these hills have flat tops, which may be another reason for describing them as maimed, as if without a sharp prominent peak they are incomplete. Another naming matter is that, since they are termed hills, we must call them so. Yet technically, using English measurements, any such over 1,000 feet or to some experts 1,500 feet, could be called a mountain. Even Ingleborough is called just a hill. Nevertheless, when one sees the great masses of Ingleborough, Whernside or

Pen-y-Ghent, they seem to deserve a more appropriate name.

Blakey Topping: Blakey is 'black mound' from OE blaec + ON haugr whilst Topping is self-explanatory. The legend is that soil scooped out by giant Wade to create the Hole of Horcum was flung away by him to make this protuberance or else Roseberry Topping, and perhaps depending on how long he worked there was not enough soil for both.

Bridestones: From OE brîd + stân. Sandstone tors about nine miles west of Scarborough.

Brimham Rocks: East of Pateley Bridge. Brimham means 'Brim's home'. Locally brim apparently means 'a high place exposed to weather'. Their scenery is so like that on Dartmoor that Sherlock Holmes's story *The Hound of the Baskervilles* was filmed by TV there.

Buckden Pike: (702 metres) Beyond Kettlewell, named from the hamlet in the valley below it. Probably from OE bucc 'stag' + OE denu 'valley' + French pique 'shaft'. For the last element compare The Langdale Pikes.

Buttertubs Pass: Connecting Muker and Hawes. The Buttertubs (OE butere + ME

tubbe) are deep holes by the roadside in the shape of old farmhouse tubs where butter was made by stirring milk hard by hand, not in a factory, a system now hardly known about by town-dwellers. The water dropping down to create the Buttertubs emerges from a cave 300 feet below.

Chevin: A steep ridge overlooking Otley. Associated with Welsh cefn 'ridge'. It makes a popular walk up from the town, rather comparable with climbing from Ilkley up to Ilkley Moor.

Cleveland Way: It stretches from Kilburn's White Horse via Saltburn to Filey Brigg or vice versa.

Caves: From Latin cavus 'a hollow'. For intrepid lovers of the sport of caving there are e.g. White Scar Cavern near Ingleton and Stump Cross Caverns on the Pateley Bridge–Grassington road.

Coverdale: Spelt exactly the same in 1202. From OE cofer + dale. It runs south-west from Middleham on the River Ure.

Devil's Arrows: From OE deô-fol + earh. Three great standing gritstones near Boroughbridge, probably from the Bronze Age and probably hauled from rocky Knaresborough about six miles away.

Dodd Fell: (667 metres) Between Hawes and Langstrothdale Chase. For its origin see the Northern dialect word dodd 'bare round hill or fell' and compare North Frisian dodd 'a heap'.

Easby Moor: On which stands the 50 foot monument to Captain Cook, discoverer of Australia. Easby means 'Êsi's farm' from ON Êsi + by.

Elmet: Name of a Celtic kingdom, origin unknown, which resisted the invasion of Angles for over 100 years till well into the 7th century. It appears in the name Sherburn-in-Elmet (North Riding) along with Barwick-in-Elmet (West Riding). The area is noted for its maypole dancing.

Ends of Dales: The mouth of a dale is called its end, though we might be tempted, if entering at the wider end, to call it the beginning. Thus e.g. Dale End near Danby. The other end is the head.

Eskdale: From British Esca with basic meaning 'water' + dale. Largest of the dales in the North Yorkshire Moors National Park, but unlike the others its waters flow from west to east.

Farndale: In the Yorkshire moors above Kirkbymoorside. From OE fearn 'fern' + dael 'val-

ley'. Although ancestors noticed most the ferns, visitors have been attracted in spring by the masses of wild daffodils along both banks of the River Dove. This has been going on for at least 50 years since the author was taken in a party of overseas people to see them.

Fat Betty or White Cross: Named from its appearance. A stone cross standing on the watershed at the head of Rosedale, probably indicating the way to Waterdale and the Esk Valley.

Filey Brigg: A narrow rock-ridge projecting half a mile into the North Sea. From Filey (probably ON fîfa 'cotton grass + OE leah 'clearing') + ON bryggja 'jetty, landing place'.

Fingay Hill: From ON thing-haufr 'assembly mound or hill'. A prominent round-topped hill near Osmotherley where a Scandinavian council used to deliberate.

Flat Howe: On the A169 1.5 miles south of Sleights. 'Flat mound' from ON flat + haugr 'hill'. It is an ancient burial ground.

Fylingdales Moor: Inland from Robin Hood's Bay. Fylingdales means 'settlement of the people of Fygele' from the name of their OE ancestor. The moor was home of the Ballistic Early Warning Station, a futuristic peace-keeping installation whose mysteries were contained in what seemed like three enormous white golfballs.

Gills: From ON gil they are ravines. Thus Collier Gill near Goathland, Crunkley Gill 'ravine by the crooked cliff' from crum + cliff + gil near Glaisdale, and Skell Gill 'temporary hut ravine' from ON skâli + gil near Aysgarth.

Glaisdale: A small valley running south-west from Egton. From a first element probably linked with Welsh glas 'blue, green' + OE dael.

Great Shunner Fell: Guarding the Buttertubs Pass means 'Sjôn's lookout hill' from ON Sjôn + ON haugr 'hill'. Being 713 metres high, it has a great claim to be included in Yorkshire's famous peaks (compare the Three Peaks Walk involving Whernside, Ingleborough and Pen-y-Ghent) because it is higher than the last-named. There is no need to 'shun' it because it is even farther away from centres of civilisation than the other three.

Holderness: 'Headland of the hold, a high-ranking officer'. A completely Scandinavian word from holdr + -er of the ON genitive singular + ON nes 'cape, headland'. A hold in the Danelaw (the area of Danish administration) could be, for instance, a military leader or powerfully great landowner.

Muker

Regarding natural features, with their attractions and drawbacks, the chief concern in Holderness has always been the constant battle between land and sea.

Hullshire: A miniature county created by Henry VI to include the parishes of Kingston-upon-Hull, Hessle, Kirk Ella and North Ferriby.

Humber Places: Here will be brought together interesting ones well-known to seafarers from places on or near the Humber's northern bank: **The Binks:** Just north-east of Spurn Point. From Northern dialect bink 'a heap'. At first they were a heap of stones. **The Canch:** Off the fishing village Paull. 'A steep sill'. Mr A.C. Binns (in providing a fine 1957 article for The Yorkshire Dialect Society) stated that he knew of no other survival of this word, but there is near Fleetwood docks, where I was for a time a docker, the Black Canch, on which vessels sometimes run aground. **The Hebble:** A stretch of water off Paull. From a dialect word hebble "to cobble, build up hastily" as a very lumpy ebb tide flows out against the wind. **Oyster Ness:** Between Brough and North Ferriby. 'Eastern headland' from ON 'east', not 'oyster', + ON nes. **Scalp:** According to the Oxford English Dictionary it is a flat-topped rock visible only at low tide, but Hedon Scalp just north-west of Paull is mud. **Skitter Ness:** From Old Norse. 'Skith's headland'. **Pudding-Pie Sand:** South of Brough in the middle of the Humber. The description is straightforward. **Whelps:** A mid stream area very impressive at times because of its waves. Seemingly named from the leaping of puppies.

Ingleborough Hill: As it is modestly called (721 metres or 2,370 feet high) means 'fortification of the Englishman' from OE Engle + burgh, referring to the Celtic ramparts still visible at the top. Its long plateau resembles from some angles that of Kinder Scout in Derbyshire, but is higher than the latter and boasts the loftiest hill fort in England.

Kilnsey Crag: A great cliff north-west of Grassington which was undercut by Wharfedale glaciers. Kilnsey is probably 'pool near the kiln' (OE cyln + sâe).

Langstrothdale: Between Dodd Fell and Pen-y-Ghent. Supposed home district of Chaucer's two Cambridge students in *The Reeve's Tale*, who came from 'far in the North I can not telle where'. Chaucer is thinking of Langstrothdale as he writes "Of one town where they born that called is Strother" and their dialect is Northern English to the extent that they use expressions like 'as clerkes say'n ('say')', an old verbal plural only recently, it seems, extinct in Lancs., and 'thou is'. But Chaucer seems deliberately vague as Langstrothdale is too isolated to have a town.

Lovely Seat: (675 metres) Just east of the Buttertubs Pass it is from OE luflîc saet. Its beauty will naturally depend on the weather.

Lyke Wake Walk: A tough 42 mile walk from near Swainby over at least seven moors (Cringle, Cold, Urra, Waterdale, Glaisdale, Egton Hill and Goathland or more if you get lost) to reach the sea about the same latitude at Ravenscar. It was invented by Bill Cowley, a great stalwart of The Yorkshire Dialect Society. The wake is a vigil over a corpse and lyke refers to the corpse itself. The dirge, an old funeral song ten verses long, was chanted by the first explorers as they trudged at night over the dreary terrain. A sample verse goes:

"When thoo frae hence away art passed,
Ivvery neet an' all,
Ti Whinny Moore thoo cums at last,

An' Christ tak up thi soul."

Comforting? It might be rather better to stay in bed and not worry about where Whinny Moore is.

Malham Cove: A 230 foot high limestone cliff eroded by a glacier. Possibly 'den among the cup-shaped hollows' from OE dative plural maelum + OE cofe 'chamber'.

Marston Moor: Eight miles west of York. Site of the 1644 Civil War battlefield. Marston originates from OE mersc 'marsh' + tûn.

Newtondale: From the North Yorkshire Moors to Pickering. From OE nîwe + tûn + dael. The private steam railway, Pickering to Grosmont, runs through it and to the east is the A169.

Penhill: (550 metres, 1,800 feet) Overlooking Wensleydale. Probably from Welsh pen 'head' + OE hyll. A rather repetitive name, but better than that of a little village in Cumbria called Torpenhow, literally 'hill-hill-hill'.

Pennines: Although pen 'head' occurs in so many placenames from Celtic, it is absent from Pennines, which was apparently a term invented by a Charles Bertram (1723-65). From before then no special word for the range has been noted.

Pennine Way: This very long

ramblers' path crosses the region covered by this book from Cowling, continues through Gargrave and Malham, climbs Pen-y-Ghent and then proceeds through Hawes, Muker, Keld, etc. on its way to Kirk Yetholme in Scotland.

Pen-y-Ghent: This time Old Celtic pen 'summit' does start the name of this famous Pennine peak (693 metres), one of the objectives in The Three Peaks Walk. E.g. from places like Settle six miles away it looks impressive, rising sharply to a point.

Pike Hill: (326 metres) Four miles west of Goathland. From French pique 'shaft' + OE hyll. Highest point on the eastern half of The North Yorkshire Moors.

Potholes: In the Dales section the Buttertubs group has been mentioned. Britain's deepest pothole, however, is Gaping Ghyll near Ingleton, from ON gapa 'to yawn, split open' + ON gill 'ravine', occasionally used as here of a swallow-hole in limestone rock. The spelling ghyll is apparently in imitation of its use by the Lakeland poet, William Wordsworth. Other well-known pots, as they are often called, are Hull Pot and Hunt Pot on the western flank of Pen-y-Ghent and Alum Pot on the

northern face of Ingleborough.

Roseberry Topping: (322 metres, 1,051 feet) Near the A173 road from Skelton to Stokesley. It is a huge conical hill looking possibly like an extinct volcano and is a favourite of people from Middlesbrough only seven miles away. Its earlier Scandinavian name was Outhenesberg 'hill of Odin' and this spectacular hill might well have been a centre for worshipping that Scandinavian god. In 1610 a fundamental change of name began with the spelling Outhenesberg or Roseberry Topping. The nearby village was often called Newton-under-Oseberry and listeners, not catching the start of that name but just the r of under, joined it to Oseberry and unintentionally created Roseberry. Topping from OE top here means 'hilltop'. These matters must of course be far nearer the truth than the intriguing legend that the giant Wade threw away preposterous loads of spare soil to result in this marvellous Topping.

Rosedale: Is named from no flower but means 'Russ's valley' from an ON by-name + dale. Starting from near Westerdale on the North York Moors this dale, irrigated by the River Seven, extends past Rosedale

Abbey to Appleton-le-Moors.

Ryedale: Extends from near Cod Beck Reservoir on the moors beyond Osmotherley to past Rievaulx Abbey. Its name comes from the River Rye (perhaps from Latin rivus 'stream') + OE dael.

Sawdondale: A small dale extending to Wykeham. 'Valley of the willows' from, rather repetitively, OE sealh 'willow' + deu 'valley' + dale.

Spurn Head or Spurn Point: Which apparently replaced the name Ravenseer Odd, is probably from a variant of Early Modern English spur 'projecting piece of land'.

Sunk Island: Is a contradiction in terms. It was originally part of the mainland and then cut off by water to make an island in the River Humber (and so marked on a 1678 map). Then the water helpfully retreated to leave it again part of the mainland. It looks as if the Humber cannot make up its mind.

Sutton Bank: Is from OE sûth + tûn + ME banke, meaning 'hill near the south homestead'. It is a notoriously steep, mile-long hill on the A170 between Thirsk and Helmsley. Motorists know it well. To meet it without warning for the first time and in darkness, as some of us have

done, can be a little nerve-racking, but in daylight a breathtaking panorama appears at the top.

Swaledale: Most northerly of the Yorkshire dales, it is from OE swelh 'whirlpool' + dael. Swaledale used to be an area of lead-mining.

Three Howes: On Farndale Moor along with another Three Howes and **Two Howes** near Goathland. All these, from ON haugr 'barrow', are ancient burial mounds.

Troutsdale: In which Hackness stands. A pleasant-sounding dale whose waters race into the Derwent. The meaning is 'Trut's valley', from the ON personal name Trûtr + ON dalr.

Urra Moor: Urra is by contrast a most unpleasant name, from OE horh 'filth' + ON haugr 'hill'. On it, at 454 metres and near Seave Green at 382 metres, are the two highest points of the Cleveland Hills.

Wharfedale: Naturally gathers its name from what helped to make it, the River Wharfe connected with ON hverfr 'winding'.

Wensleydale: Is the best-known Yorkshire dale. At 40 miles long it is the longest (countless others jostle for the less important title of shortest).

It means 'valley of Waendel's forest clearing' from his OE personal name + leah + dael. It used to be called Yuredale or Yoredale because of its River Ure, but that custom has almost gone. The popular TV programmes about James Herriot the vet were filmed in Wensleydale.

Westerdale: This 'western dale' from OE west + dael, is near the head of Eskdale.

Whernside: (736 metres) A 13th century spelling Qwernsyd suggests 'hill where millstones are quarried' from OE cweorn or ON kvern 'millstone' + OE sîde 'side'. This is the highest peak in Yorkshire and to conquer it, separately or as part of The Three Peaks Walk, as it silently broods next to the Settle-Carlisle railway, must be the determination of many a rambler. The odd thing is that its name is too deceptively modest, for there is near Kettlewell another Whernside, justifiably called Great (but tactfully not Greater) Whernside which is somewhat lower at 704 metres. The 'Three Peaks' Whernside, provided supporters of the other agree, might benefit from a little name adjustment.

Wolds Way: From the Humber Bridge to Filey (or vice versa) via Market Weighton and passing the deserted medieval village of Wharram Percy and Pocklington.

Yorkshire Wolds: From OE weald, they are the high tract of chalk hill extending in a crescent near the Humber at aptly named Wauldby 'wold settlement' and near South Cave to the North Sea at Flamborough Head. Originally the OE word meant 'high forest land' as in the Kentish Weald, but later as in Yorkshire it came to denote high open land which might have trees in parts.

Hubberholme Church

STREET NAMES

First we must answer one big question; did Watling Street run through Yorkshire? At first glance the query seems rather foolish, as records show that Watling Street was just the name for a road near St Albans written in OE times as Waeclinga straet or Waetlinga straet, which was sensible because it led to Waeclingaceaster, now St Albans but then 'the Roman fort of the Waeclings (Wacol's people)'. However, the name spread. Soon the main line of communcation from Dover through London and St Albans to Wroxeter in Shropshire was Watling Street and then it was further extended in name to Chester. Not to be outdone, other areas, including parts of Yorkshire, assumed the name. That between York and Corbridge was called Watling Street, particularly the North Riding part of the A1 north from Leeming.

Easily the best preserved stretch of Roman road in Yorkshire, 1.2 km long, is south-west of Goathland on Wheeldale Moor. It might even be older than Roman (see Wilkinson *What the Romans Did for Us*). It is called **Wade's Causeway**. However, it was not from General Wade, who built his military track through the Cairngorm mountains in Scotland, but according to legend from the giant Wade, supposed creator of Roseberry Topping, who built it for his equally giant wife to step along to milk her cows. Whoever made this road, it is probably of its type the finest in Britain.

Turning now to Yorkshire street names rather than long roads, Roman or otherwise, these can be just as fascinating as those of towns, villages and hamlets, and doubtless more so to the folk living there or connected with them. Unlike the soul-less New York street naming system (42nd Street, etc.) Yorkshire's can from time to time embody vivid history and customs.

Take streets in **Beverley** for instance. For animal names it has its Bull Ring like Birmingham and Hengate because those fowl were kept or found there. Occupations provide another source of such names. Thus we have **Walkergate**, 'street where walkers lived', not hikers but cloth-fullers; Wednesday Market and Saturday Market in which were many traders; and **Toll Gavel** (OE toll + OE gafol 'tribute, rent') where tolls were col-lected just as today in car-park toll booths. Other examples come from people living there of other regions or nationalities. Thus **Flemingate**, where people from Flanders tended quite nat-urally to live together; and **Loundress** 'Londoners' Lane', recorded in Beverley records for 1660 as the Londiners Street.

Other streets were named from what they led to, like **Keldgate**, **Ladygate** and **Lairgate**, leading respectively to a spring (ON kelda), the Church of Our Lady, and one leading to a barn (ON hlatha, from which we get the Yorkshire dialect word laithe) plus the very common Scandinavian word-ele-ment gata 'gate', not the five-barred type but a road, way or street. Of less pleasant types we find **Hellgarth Lane** from per-haps a haunted field or a place where criminals, etc. were buried, and **Lurk Lane** apparently from a dark, dirty street from which robbers might spring out at you.

In **Bridlington**, as could be expected, Scandinavian influence has been at work on the street names, in **Applegarth** 'apple orchard' Lane, which includes ON garth 'enclosure', and **Kirkgate** from ON kirkja + gata. **Hedon** helps by supplying some old occupational names. Its butcher, baker and (no, not the candlestick-maker) the shoemaker are recorded respective-ly in its **Fletcher Gate** (ME flessher), **Bakster Gate** (OE bae-cere) and **Souttergate** (ME souter). **Hornsea**'s special contri-bution, used more often of settlement or buildings than streets, is Newbegin 'new building' from ME nîwe + bigging.

Now comes **Hull**'s turn. It can rejoice in streets named from the religious mendicant orders, **Blackfriargate** from the Dominicans and **Whitefriargate** from the Carmelites according to their different garbs. Among other city names it can produce **Bowlalley Lane** from the game played there; **Dagger Lane** or in a 14th century spelling Daggardlane 'Street where daggers were made (or more grievously used)'; **Finkle Street** 'a narrow crooked street', for which compare old Northern dialect fenkl and Danish vinkel, both meaning 'corner'; **Land of Green Ginger**, explained as the name taken from an old garden used for growing pot-herbs; and most infamously **Rotten Herring Street**. This last strangely commemorates John Rotenhering, who made himself well-known around 1350. Did he manage to sell quantities of stale fish? He is also remembered by **Rotten Herring Staith** (OE staeth), a landing place named from his family.

As befits a country area, **Market Weighton** supplies an animal name, **Hungate** 'street where dogs are kept or found' from OE hund + ON gata. **Masham**'s example might also seem an animal name but is quite different. It has **Badger Lane** from small traders. Many of course give good service but its Masham sense is from the word for 'pester' in pleas like 'Stop badgering me'. **Pickering** has **Saltergate**, occupational from OE saltere 'salt-worker' for its road to Saltburn, probably very suitable for a road carrying salt mined in Cleveland.

Richmond provides a mixed assortment. Richmond Castle was given a French-influenced name and, when a town developed below the castle, the existence there of a street called **Frenchgate** (ME Frenshe + ON gata), even though the street name was half Scandinavian, suggests an alien population attracted to the security of the Norman fortification. **Bargate** (including OE bere) means 'a road along which barley was led', **Ankirkirk** (OE ancre + ON kirkja) means 'church of the anchorites', and **Newbiggen**, comparable with a Hornsea example previously given, means 'new building' (ME nîwe +

bigging). A mournful example completes the Richmond offering: **Gallowgate** 'street leading to the gallows (OE gaelga)'.

In **Ripon** like other towns few names from agriculture would be expected, but it nurtures one happy exception, **Blossomgate**. Another interesting name there is **Horse Fair**, reminding us of times when it ran its annual sale of horses.

Scarborough's unusual street names fittingly contain some with traces of the seaside. Thus along with **Carter Gate** 'carriers' way', fairly normal for north-east England by depending on Scandinavian gata 'gate' where road would elsewhere be usual, we meet **Sandgate** 'road to the sands' and **Dumple Street** from OE dumpel and dump 'deep hole in the bed of a river or pond'.

Now in the survey we reach **York**, first prize in street name collecting. Indeed it has fostered so many interesting ones that only those still existing can be concentrated upon. As a large thriving city it employed in older times many types of tradesmen, from whom we get e.g. **Blake Street** where bleaching (OE blâecan) was done, **Colliergate** with its coal-dealers (ME colyer + ON gata), metalworkers in **Coppergate** (ON koppari), the making of felt (OE felt) in **Felter Lane** and fishmongers in **Fishergate**. The **Shambles**, one of the most famous and quaint streets with buildings almost touching each other across the road, was formerly called Fleshshambles because it had benches for the sale of flesh 'meat', whilst another specialised occupation appears in the all-Scandinavian word **Skeldergate** (ON skjolde + gata) 'abode of shield-makers'.

People of other nationalities could be rather a problem and tended to live near their fellow-countrymen. Consequently the Jews were usually together in Jewbury, the Jewish quarter. Yet, when Edward I expelled Jews from the city, they settled in what had been **Bretegate** 'street of the Britons', which took the name Ju-Bretgate which gave modern **Jubbergate**.

Other miscellaneous but highly interesting street names follow alphabetically; a few streets were named from buildings, yielding The **Baile** (OFr baille 'palisade') – compare London's

Old Bailey – and the **Bedern** 'prayer-house', connected with OE biddan 'to entreat'. The city is clearly short of countryside street names, but there is the delightfully sounding **Blossom Street** which used to be even more agriculturally Plouswayngate 'street of the ploughmen'. Then there are **Bootham**, a road, and **Bootham Bar**, a gate in the city wall, near the site of booths for the ancient city market (from ON bûthum 'at the booths').

Castlegate is 'the way to the castle' while **Coney Street** is yet another Norse-influenced term and sounding important as it arises from ON konungr 'king'. Religion and clerics appear in **Gillygate** from a church dedicated to St Giles, and in **Monkgate** from OE munuc + ON gata. **Goodramgate**, a distorted street joining Monks Bar to Petergate, is from a popular ON name Guthornr. **Grope Lane** used to be a dark, narrow alley where citizens had to grope (OE grâpian) their ways among the puddles and **Micklegate** (OE mycel + ON gata) means 'great street'.

Fresh York suburbs have appeared but rarely fresh place-names, an exception being **Newbiggin Street**, with OE nîwe + and bigging 'building' as in two other street names already noted. Ease of walking was a blessing, accounting for The **Pavement** and **Stonegate** 'street paved with stones', while **Stonehow Lane** must have been occasioned by a nearby stone arch (ON stein-hogi). To conclude this most curious list comes York's strangest street name, **Whip-ma-Whop-ma-Gate**, probably the site of an old whipping-post and pillory.

In compiling this Yorkshire street name guide a little fussiness over exact meaning of names may have been noticed. This might be through having lived, for years each time, in opposite conditions – in an Abbot's Walk miles from any abbey, a Promenade Road away from the promenade, a Drury Lane not at all theatrical, a Cumberland Street never in Cumberland, a close which was quite open and a half-mile crescent which could only by a twist of imagination be that shape. Apologies therefore for bias if any. One could start wondering whether there are some folk living in squares which are actually rectangles, whether all avenues should by law be forced to have trees, and so forth; but that would be taking street name study much too far.

Far more important are the people who live there. It can happily be reported, however, that in the matter of street names and compared, say, to the New York system of naming them by numbers, street names of the North and East Ridings, old and new, look far more individual and friendly.

Hawnby

PRONOUNCING PLACENAMES

How to say a placename can occasionally be as embarrassing as getting wrong someone's personal name. Once, rather out of my element then in London, I pronounced Pall Mall as if it rhymed with hall and still feel a trifle ashamed of the laughter it caused in that particular Cockney circle. There can often be a Standard English way of saying a local name along with a different historical and dialectal way tending to make a countryman at least smile if you get it badly wrong. If in doubt, one way is to consult a pronouncing dictionary (good too for other words besides placenames) like Daniel Jones's *English Pronouncing Dictionary* or similar, and to keep in mind older pronunciations, a good many still alive. Here for guidance are some local pronunciations from the North and East Ridings:

Aldborough	Awbra	Bursea	Bossy
Amotherby	Amby + Emerby	Burstwick	Bostwig
Argam	Arram	Burton Agnes	Botton Agnes
Aysgarth	Eh-ersgarth + Ayska	Countersett	Coonterset
Barmby	Bomby	Catterick	Catrick + Catherick
Barugh	Baaf	Cherry Burton	Cherry Botton
Beswick	Bezzick	Crayke	Creherk
Beverley	Bev(er)ly	Dalby	Dawby
Birdsall	Bodsall	Dalton	Dawton
Birkby	Borkby	Easingwold	Eeazinud
Bishop Burton	Bishy Botton	Eryholme	Errium
Bridlington	Brid	Filey	Fila + Fahla
Bubwith	Bubith	Giggleswick	Gilzick

Goodmanam	Goodmadam		Weeton
Goathland	Goadland	Marske	Mask
Gowthorpe	Gawthrap	Masham	Massam
Great Ayton	Canny Yatton	Metham	Mettam + Meetham
Grosmont	Grossmont	Middlesbrough	Middlesbruf
Guisborough	Geezbra + Gizzbra	Moulton	Mohton
Gunnerside	Gunnersit	Moxby	Mohsby
Halsham	Awzam	Muker	Miooker
Haltamprice	Autumnprize	Old Malton	Awd Mawton
Hawes	Taws (= the Hawes)	Osmotherley	Ozmala
Hawsker	Oscar	Ousethorpe	Oozethrap
Hedon	Eddn	Owthorne	Oothran
Helmsley	Emzla	Raskelf	Raskil
Hessle	Ezzel	Rievaulx	Rivis
Hole of Horcum	Awcam	Roos	Rooaz + Russ + Rose
Hotham	Utham	Routh	Rooth
Howden	Ohden	Runswick	Runzick
Huggate	Uggit	Rushton Parva	Laatle Reeaston
Irton	Orton	Ruswarp	Ruzzap
Jervaulx	Jarvis + Jervoh	Saltburn	Sawtban
Jolby	Joeby	Sancton	Santon
Keyingham	Keni(n)gham	Scalby	Scawby
Kilnsea	Killsy	Sigglesthorne	Sillsthran
Kilnwick Percy	Killick Piercy	Skirpenbeck	Skopmbeck
Kirkburn	Kokburn	Skipwith	Skipith
Kirkham	Kokham	Skirlington	Skeleton
Leighton	Leeton	Sleights	Sleets
Little Weighton	Laatle Weeton	Southburn	Soothbrun
Londesbrough	Loansbra	Spaldington	Sparrerton
Malton	Mawton	Staithes	Steeaz
Market Weighton	Market	Stokesley	Stohzla

Storwood	Storrad
Swaledale	Swawdill
Thirsk	Thosk + Thrusk
Thirkleby	Throttleby
Thixendale	Thissendill
Thorngumbald	Gumbathan
Ure	Yure + Yore
Wath	rhymes with Kath
Wauldby	Wawdby
Weaverthorpe	Weirthrap
Welwick	Wellick
Wharran Percy	Warram Piercy
Whitby	Widby
Withernsea	Withransy
Worsall	Warsall
Wrassall	Razzle
Wycliffe	Wicliff
Wykeham	Wickam
Yokefleet	Yuckflit + Yokeflit
Youlthorpe	Yothrap

Finally a few Standard English pronunciations which might baffle a stranger (Standard English Pronunciation in approximate spelling):

River Alne	Awn or Ahn
Butterwick	Butterick or Butterwick
Dalby	Dawlby or Dolby
Dalton	Dawlton or Dollton
Goathland	As it is spelt, not Goatland
Guisborough	Gizzborough
Harewood	Harewood or Harwood
Malton	Mawlton or Mollton
River Ouse	Ooze
Rievaulx	Reevo
Ruswarp	Ruzzap or less often Ruswarp
Sleights	Slights

West Tanfield

INDEX

E

Earswick 15
Easby 32
Easby Moor 69
Easington 44
Easingwold 15
Eastburn 44
Eden Camp 59
Egton 33
Eller Beck 63
Ellerker 44
Ellerton 23, 44
Elloughton 44
Elmet 69
Ends of Dales 69
Esk (River) 64
Eskdale 69
Everingham 44
Everley 33

F

Faceby 33
Fadmoor 33
Fangfoss 44
Farndale 69
Fat Betty 69
Felixkirk 33
Fell Briggs 44
Felter Lane 80
Ferrensby 15

Filey 44
Filey Brigg 69
Fingay Hil 69
Finkle Street 79
Fishergate 80
Fitling 44
Flamborough 44
Flat Howe 69
Flaxton 15
Flemingate 78
Fletcher Gate 78
Flixton 44
Flotmanby 45
Folkton 44
Fordon 45
Foston 15
Foston-on-the-Wolds 45
Foxton 33
Frenchgate 79
Frodingham 45
Fulford 15
Full Sutton 45
Fylingdales Moor 69
Fylingthorpe 33

G

Gallowgate 80
Ganton 45
Gargrave 23

Gate Helmsley 15
Gilling 15
Gills 70
Gillygate 81
Glaisdale 70
Goathland 33
Goldsborough 16
Goodmanham 45
Goodramgate 81
Goole 45
Grassington 23
Great Ayton 33
Great Barugh 16
Great Shunner Fell 70
Greenhow Hill 23
Greta (River) 64
Grinton 24
Grope Lane 81
Grosmont 33
Guisborough Priory 59
Gunnerside 24

H

Habton 16
Hackness 45
Hallikeld Hall 59
Haltemprice 46
Hanging Grimston 46

Grassington

BIBLIOGRAPHY

Brook, G. L., *English Sound Changes*, Manchester University Press, revised, 1957.

Cameron, K., *English Placenames*, Book Club Associates, 3rd Edition, 1977.

Ekwall, E., *Concise Oxford Dictionary of English Placenames*, 4th Edition, 1959.

Hudson & Co., *Historic Houses and Gardens*, 2,000 Edition, Banbury.

Johnston, Rev. J. B., *Placenames of England and Wales*, John Murray, 1915.

Mills, A. D., *Oxford Dictionary of Placenames*, 2nd Edition, 1998.

Morris, R. W., *Yorkshire Through Placenames*, David and Charles, 1982.

Muir, R., *Dales of Yorkshire*, Macmillan, 1991.

Rhea, N., *Portrait of the North Yorkshire Moors*, Robert Hale, 1985.

Smith, A. H., *Placenames of the North Riding of Yorkshire*, C.U.P., 1928.

Smith, A. H., *Placenames of the East Riding of Yorkshire and York*, C. U. P., 1937.

Turner, J. H., *Yorkshire Placenames as recorded in the Yorkshire Domesday Book, 1086*, published by the author, Bingley, undated.

Wrander, N., *English Placenames in the Dative Plural*, 1983, Lund Studies in English, No.65.

Yorkshire Dialect Society, Transactions for years 1957, 1964, 1973 and 1990.